IET POWER AND ENERGY SERIES 101

Methane and Hydrogen for Energy Storage

Other volumes in this series:

Volume 1 **Power Circuit Breaker Theory and Design** C.H. Flurscheim (Editor)
Volume 4 **Industrial Microwave Heating** A.C. Metaxas and R.J. Meredith
Volume 7 **Insulators for High Voltages** J.S.T. Looms
Volume 8 **Variable Frequency AC Motor Drive Systems** D. Finney
Volume 10 **SF_6 Switchgear** H.M. Ryan and G.R. Jones
Volume 11 **Conduction and Induction Heating** E.J. Davies
Volume 13 **Statistical Techniques for High Voltage Engineering** W. Hauschild and W. Mosch
Volume 14 **Uninterruptible Power Supplies** J. Platts and J.D. St Aubyn (Editors)
Volume 15 **Digital Protection for Power Systems** A.T. Johns and S.K. Salman
Volume 16 **Electricity Economics and Planning** T.W. Berrie
Volume 18 **Vacuum Switchgear** A. Greenwood
Volume 19 **Electrical Safety: a guide to causes and prevention of hazards** J. Maxwell Adams
Volume 21 **Electricity Distribution Network Design**, 2nd Edition E. Lakervi and E.J. Holmes
Volume 22 **Artificial Intelligence Techniques in Power Systems** K. Warwick, A.O. Ekwue and R. Aggarwal (Editors)
Volume 24 **Power System Commissioning and Maintenance Practice** K. Harker
Volume 25 **Engineers' Handbook of Industrial Microwave Heating** R.J. Meredith
Volume 26 **Small Electric Motors** H. Moczala *et al.*
Volume 27 **AC-DC Power System Analysis** J. Arrillaga and B.C. Smith
Volume 29 **High Voltage Direct Current Transmission**, 2nd Edition J. Arrillaga
Volume 30 **Flexible AC Transmission Systems (FACTS)** Y-H. Song (Editor)
Volume 31 **Embedded generation** N. Jenkins *et al.*
Volume 32 **High Voltage Engineering and Testing**, 2nd Edition H.M. Ryan (Editor)
Volume 33 **Overvoltage Protection of Low-Voltage Systems**, Revised Edition P. Hasse
Volume 36 **Voltage Quality in Electrical Power Systems** J. Schlabbach *et al.*
Volume 37 **Electrical Steels for Rotating Machines** P. Beckley
Volume 38 **The Electric Car: Development and future of battery, hybrid and fuel-cell cars** M. Westbrook
Volume 39 **Power Systems Electromagnetic Transients Simulation** J. Arrillaga and N. Watson
Volume 40 **Advances in High Voltage Engineering** M. Haddad and D. Warne
Volume 41 **Electrical Operation of Electrostatic Precipitators** K. Parker
Volume 43 **Thermal Power Plant Simulation and Control** D. Flynn
Volume 44 **Economic Evaluation of Projects in the Electricity Supply Industry** H. Khatib
Volume 45 **Propulsion Systems for Hybrid Vehicles** J. Miller
Volume 46 **Distribution Switchgear** S. Stewart
Volume 47 **Protection of Electricity Distribution Networks**, 2nd Edition J. Gers and E. Holmes
Volume 48 **Wood Pole Overhead Lines** B. Wareing
Volume 49 **Electric Fuses**, 3rd Edition A. Wright and G. Newbery
Volume 50 **Wind Power Integration: Connection and system operational aspects** B. Fox *et al.*
Volume 51 **Short Circuit Currents** J. Schlabbach
Volume 52 **Nuclear Power** J. Wood
Volume 53 **Condition Assessment of High Voltage Insulation in Power System Equipment** R.E. James and Q. Su
Volume 55 **Local Energy: Distributed generation of heat and power** J. Wood
Volume 56 **Condition Monitoring of Rotating Electrical Machines** P. Tavner, L. Ran, J. Penman and H. Sedding
Volume 57 **The Control Techniques Drives and Controls Handbook**, 2nd Edition B. Drury
Volume 58 **Lightning Protection** V. Cooray (Editor)
Volume 59 **Ultracapacitor Applications** J.M. Miller
Volume 62 **Lightning Electromagnetics** V. Cooray
Volume 63 **Energy Storage for Power Systems**, 2nd Edition A. Ter-Gazarian
Volume 65 **Protection of Electricity Distribution Networks**, 3rd Edition J. Gers
Volume 66 **High Voltage Engineering Testing**, 3rd Edition H. Ryan (Editor)
Volume 67 **Multicore Simulation of Power System Transients** F.M. Uriate
Volume 68 **Distribution System Analysis and Automation** J. Gers
Volume 69 **The Lightening Flash**, 2nd Edition V. Cooray (Editor)
Volume 70 **Economic Evaluation of Projects in the Electricity Supply Industry**, 3rd Edition H. Khatib
Volume 72 **Control Circuits in Power Electronics: Practical issues in design and implementation** M. Castilla (Editor)
Volume 74 **Power Electronic Converters and Systems: Frontiers and applications** A. M. Trzynadlowski (Editor)
Volume 75 **Power Distribution Automation** B. Das (Editor)
Volume 76 **Power System Stability: Modelling, analysis and control** B. Om P. Malik
Volume 78 **Numerical Analysis of Power System Transients and Dynamics** A. Ametani (Editor)
Volume 79 **Vehicle-to-Grid: Linking electric vehicles to the smart grid** J. Lu and J. Hossain (Editors)
Volume 86 **Advances in Power System Modelling, Control and Stability Analysis** F. Milano (Editor)
Volume 905 **Power system protection**, 4 volumes

Methane and Hydrogen for Energy Storage

Edited by
Rupp Carriveau
and David S-K. Ting

The Institution of Engineering and Technology

Published by The Institution of Engineering and Technology, London, United Kingdom

The Institution of Engineering and Technology is registered as a Charity in England & Wales (no. 211014) and Scotland (no. SC038698).

First published 2016

The Institution of Engineering and Technology
Michael Faraday House
Six Hills Way, Stevenage
Herts, SG1 2AY, United Kingdom

www.theiet.org

British Library Cataloguing in Publication Data
A catalogue record for this product is available from the British Library

ISBN 978-1-78561-193-3 (hardback)
ISBN 978-1-78561-194-0 (PDF)

Typeset in India by MPS Limited

Contents

Preface

Commercial energy storage has moved from the margins to the mainstream as it fosters flexibility in our smarter, increasingly integrated energy systems. The energy density, availability, and relatively clean fossil profile of natural gas ensure its critical role as a fuel for heating and electricity generation. As a transportation fuel, natural gas continues to increase its market penetration; much of this has been enabled by emerging developments in storage technology. Liquefied storage of natural gas is enabling the cleanest and quietest fleet of marine ferries and barges in history. Innovative adsorbed natural gas storage technologies are facilitating lower pressure, lower cost fuel storage for the light duty transport sector. Natural gas has been identified by many as the fuel to take us to the no-carbon horizon; where a hydrogen economy waits on development. These two actors are already connected in precursor applications as transitional solutions for hydrogen handling and transportation are sought ahead of a fully established hydrogen infrastructure. This monograph explores some of leading advances in methane and hydrogen storage as well as the interesting link between these two important elements in our evolving energy system mosaic.

The monograph commences with Hydrogen Absorption for Storage by Makridis in Chapter 1. In this chapter, absorption is considered to facilitate the potential utilization of natural gas infrastructure as a hydrogen storage and delivery mechanism. This is followed by Power-to-Gas for Energy System Integration and Storage in Chapter 2, where Walker *et al.* detail the use of surplus electricity from multiple sources to generate hydrogen gas that would also be injected into the natural gas infrastructure. These two chapters highlight some of the critical links that can be made between hydrogen and natural gas to help drive a more realistic march to a carbon-free economy.

Converting carbon dioxide back into methane can present a double benefit, as it captures undesirable emissions and converts them into useful fuel. On this, Kolb *et al.* tackle the challenge of the relatively low efficiency of conventional methanation with the latest technologies from a comprehensive analysis of the literature in Chapter 3. It is with minimal modification that hydrogen can be injected into natural gas infrastructure at low concentrations. This enablement builds great potential for storage of gas produced by surplus electricity. The volumes and time independency of such storage is attractive for energy utilities. This is echoed by Abanades in Chapter 4, Production of Hydrogen from Methane Decarbonization into Power-to-Gas Scenarios.

To have a viable Power-to-Gas system in practice, stakeholders at both the supply and demand ends have to work together. The flow of power must be carefully regulated in the time and manner of its delivery. In Chapter 5, Mukherjee *et al.* carry out a simple techno-economic feasibility study of a Power-to-Gas energy hub that offers said load regulation capabilities in what is termed the ancillary services market.

To overcome the high pressure required in traditional compressed natural gas (CNG) storage, Adsorption of Natural Gas (ANG) in Metal-Organic Frameworks (MOFs) has been revealed as a promising alternative. Storage pressures are up to seven times less than those required by CNG. Such pressure reductions can lower the costs of compression and increase flexibility for storage container designs. Makal delineates how MOFs' high surface areas and chemically-tunable structures have made them prominent sorbent materials for enhancing the natural gas storage capacity of on-board vehicular fuel systems in Chapter 6.

Recently, conformable storage ANG tanks have been proposed to better utilize the limited space on vehicles. To make ANG competitive, further improvement in the high-performance sorbents like MOFs is needed. As importantly, proper thermal management is a must when it comes to reasonably short refueling times. This is unveiled thermodynamically in Chapter 7 by Ahmadi-Baloutaki *et al.*

Where would one harvest for clean methane years from now? From the large amount of gas hydrates naturally stored under our feet, according to Goyal *et al.* in Chapter 8. The naturally occurring gas hydrates contain a large amount of methane per molecule of hydrate. This is not lost on countries like the United States, India, and China that are committing significant levels of research investment to find feasible ways to economically crop and harvest methane from these hydrates.

Rupp Carriveau & David S-K. Ting

Chapter 1

Hydrogen storage and compression

Sofoklis S. Makridis[1,2]

1.1 Towards a hydrogen economy

Energy has always been the driving force in the technological and economic development of societies. The consumption of a significant amount of energy is required to provide basic living conditions of developed countries (heating, transportation, lighting, etc.). Today's energy supply has a considerable impact on the environment, since it is fuelled by the burning of fossil fuels. In addition to this, the fossil fuel reserves are decreasing while the demand for energy is rapidly rising. Climate change, the depletion and geographical segregation of fossil fuel resources, health related issues as well as energy poverty constitute the driving forces towards the pursuit of alternative energy sources. In addition, countries with no access to oil reserves are being dependent from other countries for their energy supply, with a strong impact on politics and financial issues. But apart from occasional financial recessions, the long-term need for increasing amounts of energy as countries develop will become a major rate limiting step in the growth of the world economy [1]. The last years there is an on-going research on alternative fuels in order to overcome the fossil energy dependence and to provide a sustainable growth of economies and societies.

In view of the above, countries that release the largest amounts of greenhouse gases to the atmosphere compared to the energy production are expected to minimise CO_2 emission and at the same time improve the share of "clean" energy in total energy consumption. The renewable, non-conventional energy sources, such as solar and wind energy, will remain available for infinite period. But due to the inherent nature of renewable energy resources being intermittent, there is a need to store any surplus electrical energy produced in order to be used during high energy

[1]Department of Mechanical Engineering, University of Western Macedonia, GR50132 Kozani, Greece
[2]Department of Environmental and Natural Resources Management, University of Patras, GR30100 Agrinio, Greece

demand periods [2, 3]. The following are the solutions to store surplus energy produced (either in full operation or in an experimental mode) [4–7]:

a. compressed air energy storage
b. batteries
c. flywheel (mechanical inertia) energy storage
d. hydroelectricity (pumped water energy storage)
e. superconducting magnetic energy storage
f. thermal energy storage
g. hydrogen production and then storage or injection into natural gas grid (power to gas)

In this study, only the solutions related to hydrogen are discussed.

As it is derived from the above, hydrogen is an energy carrier and it is related to a process that begins and ends with plain water, known as the "hydrogen fuel cycle." In a more particular manner, water is split into hydrogen and oxygen by the process of electrolysis and then they re-combine, producing electricity and water vapour, using a fuel cell. Furthermore, hydrogen is abundant (e.g. within water) and evenly distributed throughout the world providing security in energy. But even in this case, electricity is needed. Conventional sources of electricity are being used for electrolysis, and as a result the hydrogen carbon footprint remains more or less high. Thus, with renewable energy powered electrolysers (i.e. from solar or aeolian generators), "green" hydrogen can be produced [3, 8]. But it is important to realise that hydrogen is not a fuel source; it is an energy carrier, and has led to a worldwide development effort for hydrogen technology to power industrial, residential and transportation infrastructure – a concept known as the Hydrogen Economy [9].

On Earth hydrogen is rarely found in the pure form, but usually in a wide variety of inorganic and organic chemical compounds, the most common being water (H_2O). Hydrogen forms chemical compounds (hydrides) with nearly all other elements. Due to their ability to form long chains and complex molecules, combinations with carbon play a key role for organic life (hydrocarbons, carbohydrate) [10, 11].

The diatomic H_2 gas molecule can be produced from various sources. Currently, the most widespread process for the generation of hydrogen is the steam reforming from light carbohydrates, which, however, additionally generates undesirable CO_2 [12]. Alternatively, hydrogen can be obtained from water dissociation during electrolysis. This "green" hydrogen production from sustainable energy sources is a fully reversible process and it is a solution for sustainable ongoing hydrogen production for industries, storing "green" energy without increasing its carbon footprint and supplying energy for "green" mobility (transportation) [13].

In a Hydrogen Economy the lightest of all gases has to be processed like any other market commodity. It has to be packaged, transported by surface vehicles or pipelines, stored and transferred to the end user. There it can be converted back into electricity by fuel cells or other conversion devices. High-grade electricity from renewable or nuclear sources is needed not only to generate hydrogen by electrolysis but also for almost all essential marketing stages. At the end of the chain much less electricity becomes available than that has been invested upstream in the process of generation and marketing hydrogen [14, 15].

1.2 Hydrogen – Thermo-physical properties

Hydrogen is a chemical element with a chemical symbol H and atomic number 1. In normal conditions it is colourless, odourless and insipid gas, formed by diatomic molecules, H_2. It is the lightest element on the periodic table and the most abundant chemical substance in the universe. Hydrogen gas was first artificially produced in the early sixteenth century, via the mixing of metals with acids. English chemist Henry Cavendish was the first to identify the properties of hydrogen after he evolved hydrogen gas by reacting zinc metal with hydrochloric acid, in 1766 [16]. Seven years later, Antoine Lavoisier gave it the name "hydrogen," from the Greek words "*ὕδωρ*" and "*γίγνομαι*," which means "water-former" from its property to produce water when combusted [17].

Hydrogen consists of three isotopes, most of which is ^{1}H. The ordinary isotope of hydrogen, *H*, is known as *protium*. In 1932, Urey announced the discovery of a stable isotope, deuterium (^{2}H *or D*) with an atomic weight of 2. Deuterium is present in natural hydrogen to the extent of 0.015%. Two years later, an unstable isotope, tritium (^{3}H), with an atomic weight of 3 was discovered [18]. At normal conditions, hydrogen is in gaseous state. It has the second lowest boiling point and melting points of all substances, second only to helium. Hydrogen is a liquid below its boiling point of −253 °C (20 K) and a solid below its melting point of −259 °C (14 K) and atmospheric pressure. A phase diagram of hydrogen is shown in Figure 1.1. The boiling point of a fuel is a critical parameter since it defines the temperature to which it must be cooled in order to store and use it as a liquid [19].

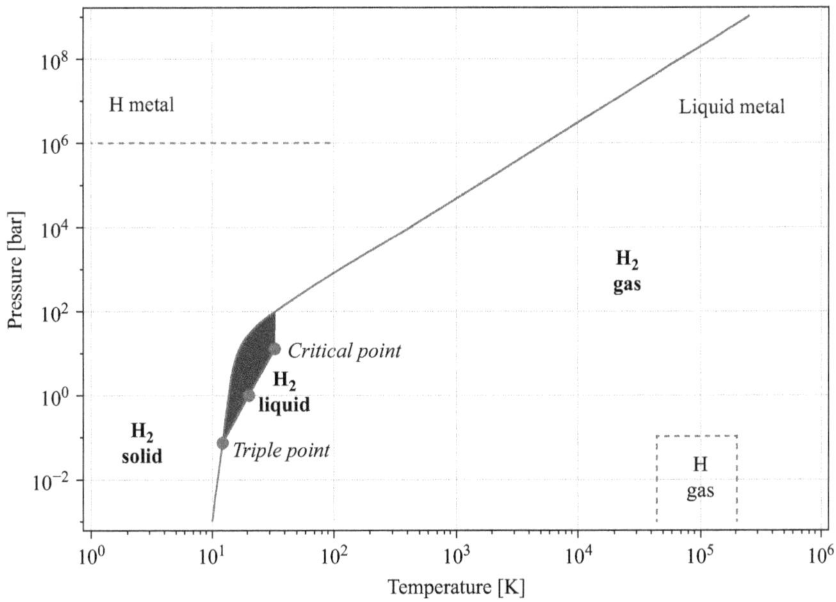

Figure 1.1 Simple phase diagram for hydrogen [20]

Table 1.1 Density of fuels in gaseous and liquid state [21]

Substance	**Gas at 20 °C, 1 bar (kg/m^3)**	**Liquid at boiling point, 1 bar (kg/m^3)**
Hydrogen	0.09	70.8
Methane	0.65	422
Gasoline	4.4	700

Hydrogen has a molecular weight of 2.01594 g and a density of 0.08988 g/l at 0 °C and 1 bar, i.e. 7% of the density of the air. For comparison, the density of typical fuels is shown in Table 1.1. Density is measured as the amount of mass contained per unit volume. Density values only have meaning at a specified temperature and pressure since both of these parameters affect the compactness of the molecular arrangement, especially in a gas. At normal conditions, gaseous hydrogen is about 8 times less dense than methane, while in liquid state it is 6 times less dense than liquid methane and 55 times less dense than gasoline. Moreover, the volume ratio between hydrogen at 1 bar and compressed hydrogen at 700 bar is 440. Finally, it is interesting to note that more hydrogen is contained in a given volume of water or gasoline than in pure liquid hydrogen (111 kg/m^3, 84 kg/m^3 and 71 kg/m^3, respectively) [21].

The density of hydrogen at elevated pressure can be estimated using the principles of thermodynamics. While the behaviour of most gases can be approximated with a high accuracy by the simple equation of state of an ideal gas ($PV = nRT$), that relates the pressure, the volume and the temperature of a given substance, the behaviour of hydrogen deviates significantly from the predictions of the ideal gas model. The resulting deviation from the ideal gas law is always in the form of expansion – the gas occupies more space than the ideal gas law predicts. One of the simplest ways of correcting for this additional compression is through the addition of a compressibility factor, designated by the symbol Z. Compressibility factors are derived from data obtained through experimentation and depend on temperature, pressure and the nature of the gas. The Z factor is then used as a multiplier to adjust the ideal gas law to fit actual gas behaviour as follows [10]:

$$PV = nZRT \tag{1.1}$$

By reducing the pressure P to the critical pressure P_{cr} and the temperature T to the critical temperature T_{cr}, a generalised compressibility factor for all gases, can be drawn as a function of $P_R = P/P_{cr}$ and $T_R = T/T_{cr}$. The value of compressibility factor Z for hydrogen at high pressures and low temperatures in Figure 1.2 shows that, at ambient temperature, a value of 1.2 is reached at 300 bar, and at low temperatures even earlier. This means that a calculation of the hydrogen mass in a container from a measurement of temperature and pressure using the ideal gas equation will result in a mass 20% greater than in reality [10].

Energy is needed to compress gases and the compression work depends on the thermodynamically compression process, as well as on the nature of the gas. This is presented by the comparison of hydrogen with helium and methane in Figure 1.3.

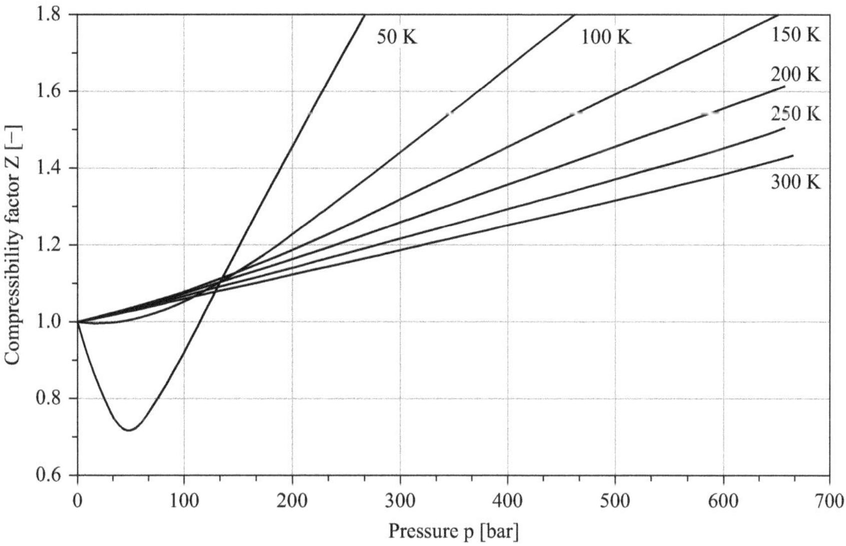

Figure 1.2 Compressibility factor Z of hydrogen [10]

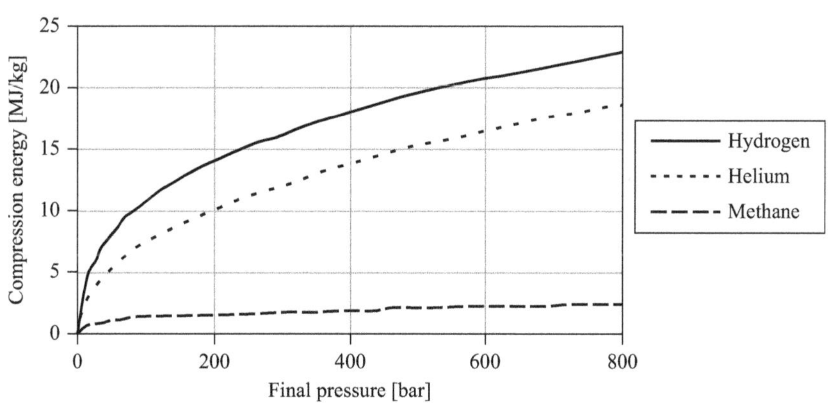

Figure 1.3 Adiabatic compression work for hydrogen, helium and methane [22]

The volumetric higher heating value (HHV; heat of formation) energy densities of different energy carrier options are illustrated in Figure 1.4. At any pressure, hydrogen gas clearly carries less energy per volume than methane (representing natural gas), methanol, propane or octane (representing gasoline). At 800 bar pressure, gaseous hydrogen reaches the volumetric density of liquid hydrogen. Therefore, compressed hydrogen cannot have the density of liquid hydrogen under any practicably achievable pressure conditions. But at any pressure, the volumetric energy density of methane gas exceeds that of hydrogen gas by a factor of 3.2. Furthermore,

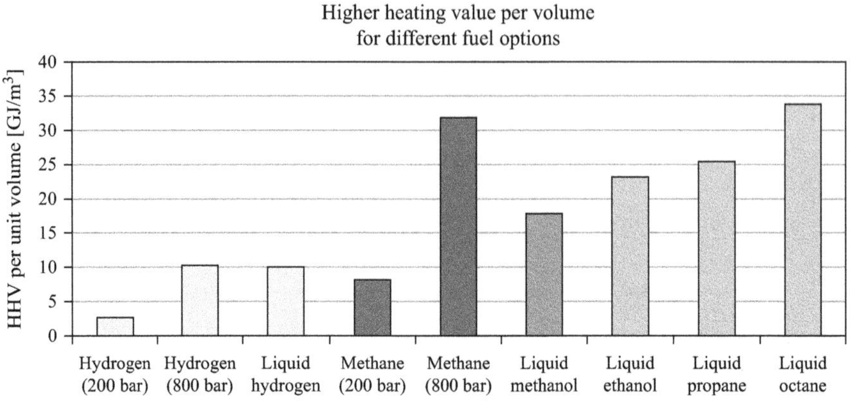

Figure 1.4 Volumetric HHV energy density of different fuels [22]

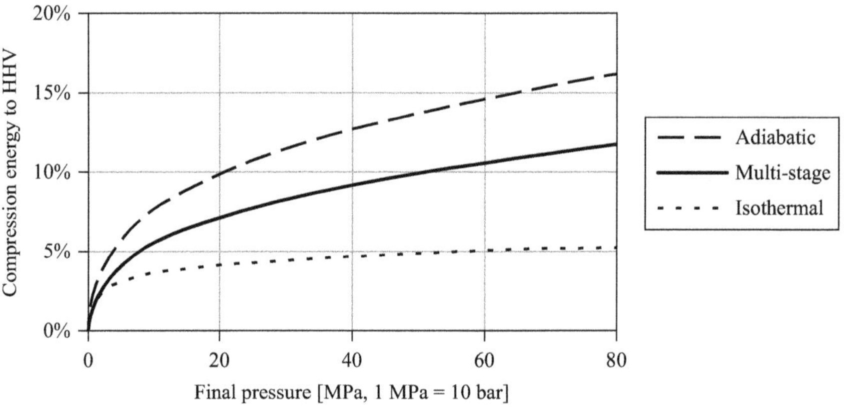

Figure 1.5 Energy required for the compression of hydrogen compared to its higher heating value, HHV [22]

at 800 bar or in the liquid state, hydrogen must be contained in specially designed pressure tanks or cryogenic containers, while the liquid fuels are kept under atmospheric pressure conditions in simple containers [22].

Generally, the compressed gas has to be cooled down after each stage to make compression less adiabatic and more isothermal. Thus, hydrogen typically is compressed in several stages. The difference between adiabatic and isothermal ideal-gas compression of hydrogen is presented as in Figure 1.5. Multi-stage compressors with intercoolers operate between these two limiting curves. Also, hydrogen easily passes compression heat to cooler walls, thereby approaching isothermal conditions. It is worth to mention that, for a final pressure of 800 bar, the compression energy requirements would amount to about 13% of the energy content of hydrogen [22].

1.3 Hydrogen storage

Hydrogen can be stored using the following three basic storage methods [23]:

a. **Compressed hydrogen gas (CGH_2)**
b. **Liquid hydrogen (LH_2)**
c. **Solid storage of hydrogen (SSH_2)**

More specifically, since the first two methods can be applied by modifying hydrogen's physical state in gaseous or liquid form, the third method can be divided into the following categories:

- Physisorption in porous materials
- Absorbed on interstitial sites in a host metal
- Complex compounds
- Metals and complexes with water

1.3.1 Compressed hydrogen gas (CGH_2)

The most common storage system is high pressure gas steel cylinders, which are operated at a maximum pressure of 200 bar. But depending on the tensile strength of the cylinder material, higher pressures can be reached. New lightweight composite cylinders have been developed that are able to withstand pressures up to 800 bar, so that hydrogen can be reach a volumetric density of 36 kg/m^3, almost half as much in its liquid form at normal boiling point. The volumetric density of n-H_2 (normal hydrogen) as a function of the pressure at three different temperatures can be displayed as in Figure 1.6 [11]. It can be easily observed that hydrogen density does not follow a linear function over the increase of pressure. A hydrogen density of 20 kg/m^3 is reached at 300 bar. The volumetric density can be increased to around 40 up to 70 kg/m^3 by compressing the gas to a pressure of up to 700 or 2000 bar, respectively. However, 2000 bar is technically not feasible [25].

1.3.2 Liquid hydrogen (LH_2)

The second method to increase the volumetric density of hydrogen is, to decrease the temperature of the gas at a constant pressure to obtain the liquid phase. Liquid hydrogen is stored in cryogenic tanks at −252 °C (21 K) at ambient pressure. Because of the low critical temperature of hydrogen of −240 °C (33 K) (Figure 1.7), the liquid form can only be stored in open systems, as there is no liquid phase existent above the critical temperature, with simultaneous small quantity losses. The pressure in a closed storage system at room temperature could increase to 10^4 bar. The simplest hydrogen liquefaction cycle is the Linde cycle, based on the Joule–Thomson effect [20].

In the Joule–Thomson effect, there is a coefficient, called Joule–Thomson coefficient, describes the extent and direction of the temperature change for an isenthalpic change of state (constant enthalpy h). The temperature that defines the transition from a positive to a negative Joule–Thomson coefficient (or the opposite) is called inversion temperature. A positive Joule–Thomson coefficient means that a

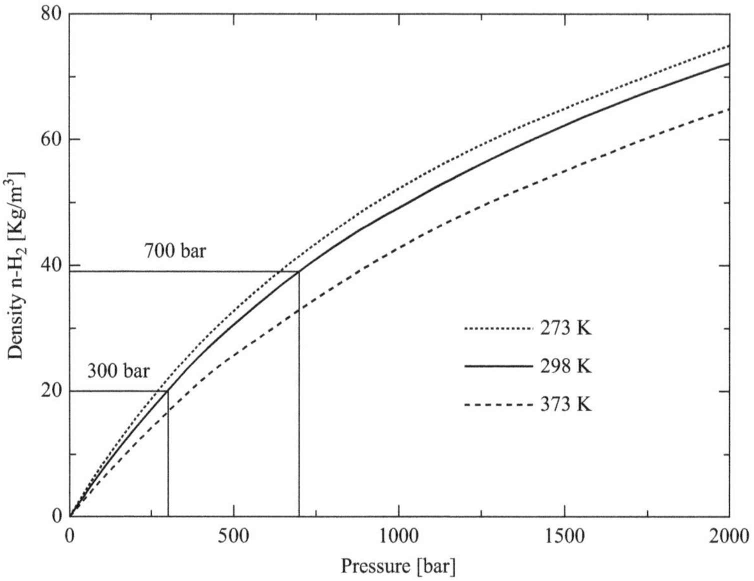

Figure 1.6 Evolution of the volumetric density of n-H_2 as a function of pressure at three different temperatures [26]

decrease in temperature takes place along an isenthalpic pressure decrease, while a negative Joule–Thomson coefficient means that an increase in temperature takes place along an isenthalpic pressure decrease. Thus, the Joule–Thomson effect occurs when a gas or gas mixture experiences a change in temperature during an isenthalpic pressure change. It is worth to mention that an ideal gas does not show any Joule–Thomson effect.

With ideal gases, the internal energy u and thus also the enthalpy h are only a function of temperature T. Thus ideal gases do not experience a change in temperature while the enthalpy remains constant, e.g. in a flow through a restriction. This means that the Joule–Thomson coefficient is zero. Hence, it can easily be judged from the T–s diagram (Figure 1.7) that, hydrogen can be regarded as ideal gas in an area where the isenthalpic lines are horizontal.

Hydrogen has a peculiar Joule–Thomson effect inversion temperature of around −83 °C (190 K) at a pressure below 50 bar. Consequently, a process other than a Joule–Thomson expansion has to be used to cool hydrogen in the range of 26.85 to −83.15 °C (300–190 K) [10]. For the liquefaction of hydrogen, the isenthalpic process of pressure decrease (throttling) is used. A throttling process proceeds along a constant-enthalpy curve (Figure 1.7) in the direction of decreasing pressure, which means that the process occurs from left to right on a T–s diagram. If hydrogen is found in the region of liquid + gas phase after the throttling (region of horizontal isobaric lines in the T–s diagram), part of it becomes liquid. The whole process is like this: hydrogen is first compressed and then cooled (under constant pressure) in a heat exchanger, before it passes through a throttle valve

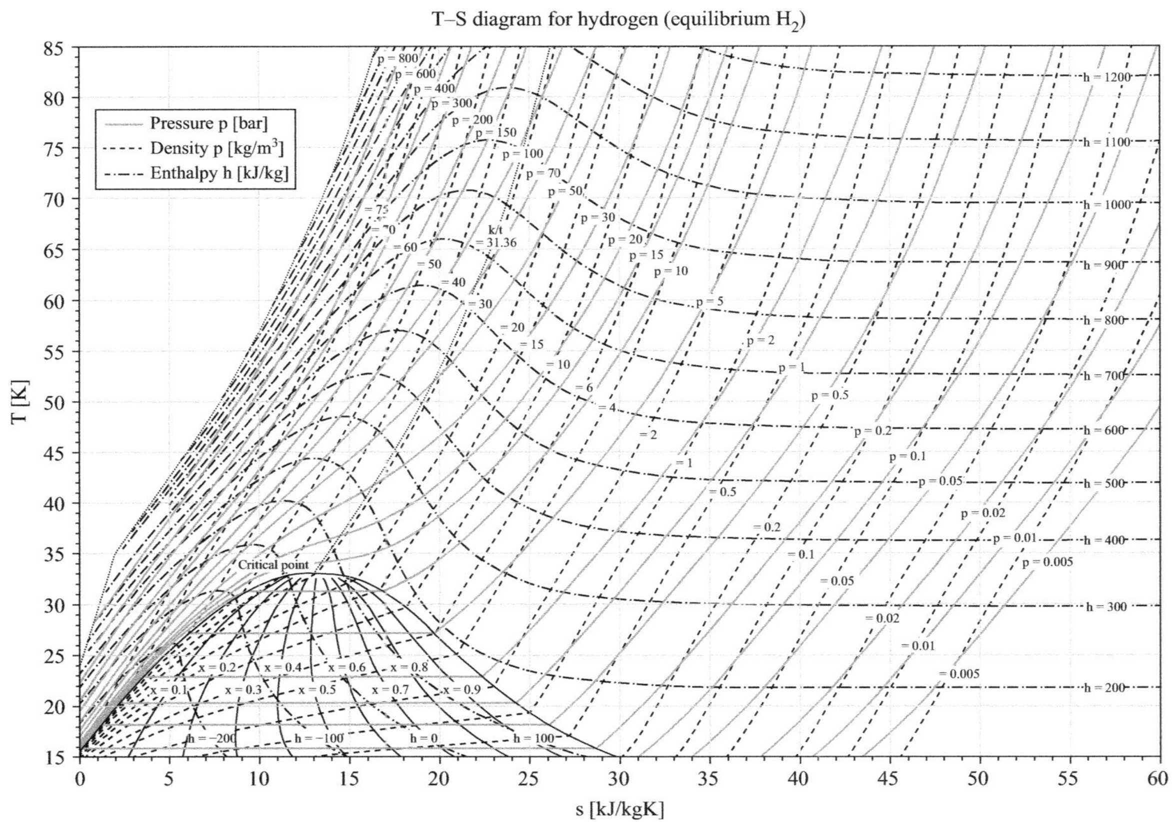

Figure 1.7 T–s diagram for equilibrium hydrogen for temperatures from −258 to −188 °C (15–85 K) [27]

where it undergoes an isenthalpic Joule–Thomson expansion, producing *some* liquid. The cooled gas is then separated from the liquid and returned to the compressor to undergo the same process.

In general, hydrogen liquefaction consumes large amounts of energy for achieving low temperatures and high pressures, as well as storage of liquid hydrogen for maintaining low temperatures at atmospheric pressure. The energy consumption, leakage along with safety issues led this solution to be applied only for special applications.

1.3.3 Solid storage of hydrogen (SSH_2)

As mentioned earlier, there are technologies that permit to store hydrogen directly by modifying its physical state into gaseous or liquid form. The traditional hydrogen storage facilities are complicated and moreover, liquid hydrogen requires the addition of a refrigeration unit to maintain a cryogenic state, thus adding weight and energy costs, and a resultant 40% loss in energy content [24, 28].

In the solid state storage there are various types of potential hydrogen storage materials. But as in every technological solution, there are always advantages and drawbacks, and the selection of a technology depends on the nature of the problem in a way that the disadvantages have less impact in the overall application.

Storage by absorption as chemical compounds or by adsorption on carbon materials has definite advantages from the safety perspective such that some form of conversion or energy input is required to release the hydrogen for use. A great effort has been made on new hydrogen storage systems, including metal, chemical or complex hydrides and carbon nanostructures [24].

Metal hydrides store atomic hydrogen in the bulk of the material. In the case of interstitial metal hydrides, the molecular hydrogen in the gas is split into atomic hydrogen on the surface of the material and then, it diffuses into the atomic structure of the host metal. In these hydrides, hydrogen acts as a metal and forms a metallic bond [29]. A large number of different metallic compounds exist that will absorb hydrogen in this manner. In most cases, however, the absorption does not occur at moderate temperature and hydrogen pressure for practical storage purposes and, the low mass of the absorbed hydrogen is only a small fraction of the mass of the host metal.

As hydrides are formed by direct reaction with gaseous hydrogen, it is important first to present the thermodynamics of hydride formation. The science and technology of reversible metal hydrides, in other words, the hydriding and dehydriding (H/D) of metals (M) by both direct dissociative chemisorption of H_2 gas (1) and electrochemical (2) splitting of H_2O are very simple. Many metals and alloys react reversibly with hydrogen to form metal hydrides according to the following reaction:

$$M + \frac{x}{2}H_2 \leftrightarrow MH_x + Q \tag{1.2}$$

$$M + \frac{x}{2}H_2O + \frac{x}{2}e^- \leftrightarrow MH_x + \frac{x}{2}OH^- \tag{1.3}$$

where M refers to a metal, a solid solution or an intermetallic compound, and Q to the heat of reaction. Since the entropy of the hydride is lowered compared to the metal and the gaseous hydrogen phase, the hydride formation is exothermic and the reverse reaction of hydrogen release is accordingly endothermic. Therefore, for hydrogen release/desorption heat supply is required.

There are several ways to show H-capacity. The reversible capacity, $\Delta(H/M)_r$, is conservatively defined as the plateau width, which can be considerably less than the maximum capacity, $(H/M)_{max}$. In practice, depending on available pressure and temperature ranges, engineering capacity is usually somewhere between $\Delta(H/M)$ and $(H/M)_{max}$. Capacity can be listed in either atomic H/M ratio or weight percent, both of which are used in some of the tables below. In calculating wt%, both H and M (i.e. not only M) are included in the denominator. In addition, it is sometimes useful to express capacity in volumetric terms, e.g. number of H atoms per unit volume (such as crystal cm^3). This measure is listed in some of the tables below as $\Delta NH/V$, where ΔNH represents the reversible capacity. Note that, this measure represents the volumetric density in crystal terms and does not include the void volumes inherent in engineering containers. In general, the mid-desorption plateau pressure, Pd, will be used in the graphs and tables below. Of course thermodynamics dictate the plateau pressures P must increase with temperature, usually close enough to the van't Hoff equation for engineering and comparison purposes,

$$\ln P = \frac{\Delta H^o}{RT} - \frac{\Delta S^o}{R} \tag{1.4}$$

where, ΔH^o and ΔS^o are the enthalpy and entropy changes of the hydriding reaction (1), T is absolute temperature and R is the gas constant. For all of the hydrides to be discussed, ΔH^o and ΔS^o are negative, i.e. the hydriding (absorption) reaction is exothermic and the dehydriding (desorption) reaction is endothermic. The knowledge of ΔH^o especially is important to the heat management required for practical engineering devices and is a fundamental measure of the M–H bond strength.

The thermodynamic aspects of hydride formation from gaseous hydrogen are described by Pressure-Composition Isotherms (PCI), as presented in Figure 1.8. When solid solution and hydride phases coexist, there is a plateau in the isotherms, the length of which determines the amount of hydrogen stored. In the pure β-phase, the hydrogen pressure rises steeply with the concentration. The two-phase region ends in a critical point, a critical temperature noted as T_c, above which the transition from the α- to β-phase is continuous.

The value of the enthalpy is an index of stability of a metal hydride. The higher the absolute value of the enthalpy shows a high degree of stability of the hydride, low dissociation pressure and the requirement of rather high temperatures to decompose it in order to liberate hydrogen [20].

At low concentration, as hydrogen initially diffuses into the pure metal interstices, the equilibrium pressure rises sharply, and a solid solution phase (α-phase) is formed. The α-phase has the same crystal structure as the bare metal. After the physical sites are saturated, a second phase, the metal hydride, begins to form.

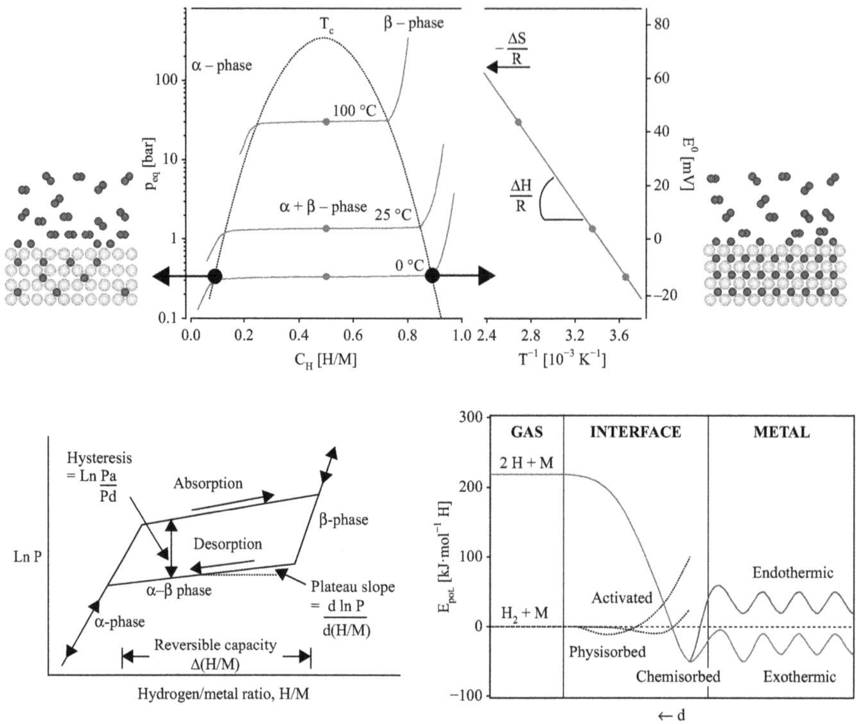

Figure 1.8 Schematic Pressure-Composition Isotherm, and van't Hoff plot giving the enthalpy of hydride formation (up). α is the solid solution of hydrogen and β is the hydride phase [20]. Full hydrogen absorption-desorption cycle (down-left) and Potential energies of $2H + M$ and $H_2 + M$ reactions at a gas-metal interface (down-right)

At this point, hydrogen is distributed methodically or randomly (depending on the hydride) in the metal host lattice and the concentration varies slowly with temperature. In this two-phase region the concentration increases while the hydrogen pressure is (ideally) constant [10]. Thus, a plateau is observed on the pressure versus hydrogen/metal ratio for fixed temperature. When the reaction is complete, another sharp pressure rise is observed if more hydrogen is added. At this point, nucleation of a higher concentration phase (β-phase) occurs. As the hydrogen pressure increases, the concentration also increases until the attractive *H–H* interaction becomes important [30]. The system now has three phases (α, β and hydrogen gas) and two components (metal and hydrogen). At a given temperature, the plateau pressure represents one point on a pressure-temperature (van't Hoff) Plot.

For practical purposes metal-hydrides are intermetallic compounds, which when exposed to hydrogen gas at certain temperatures and pressures, they absorb large quantities of hydrogen gas forming hydride compounds. The formed hydrides can then, under certain temperatures and pressures, desorb the stored hydrogen.

Hydrogen is absorbed interstitially in the metal lattice expanding the parent compound in the atomic and macroscopic level due to the Potential energies at a gas-metal interface. Such metal-hydrides can expand as much as 30% causing the decrepitation of the original ingots into fine powders. Metal hydrides represent an exciting method of storing hydrogen. They are inherently safer than compressed gas or liquid hydrogen and have a higher volumetric hydrogen storage capacity. Some hydrides can actually store hydrogen in densities twice as much of that of the liquid hydrogen (0.07 g/cm^3).

There are a number of important hydride properties that must be considered in addition to the primary PCT properties. Some of the more important ones are listed below.

Activation is the procedure needed to hydride a metal the first time and bring it up to maximum H-capacity and hydriding/dehydriding kinetics. The ease of initial H_2 penetration depends on surface structures and barriers, such as the dissociation catalytic species and the oxide films. A second stage of activation involves internal cracking of metal particles to increase reaction specific surface. Decrepitation means the self-pulverisation of large metal ingots and large particles into powder (mostly single grain fine particles). This phenomenon results from a combination of hydriding volume change and the brittle nature of hydriding alloys. The morphology of the decrepitated powder affects heat transfer and also the tendency of powder migration into undesirable places in the hydrogen storage tank, like valve seats, assisting pipes. Unfortunately, most hydride powders have poor heat transfer coefficients and require engineering means for thermal enhancement (e.g. Al foam, internal fins). The morphology of the power can affect packing, which in turn can lead to internal gas impedance and container deformation.

Kinetics of hydriding and dehydriding can vary markedly from alloy to alloy. Fortunately, many room temperature hydrides have excellent intrinsic kinetics, so that the cycling of storage containers tends to be limited by heat transfer designs or accidental surface contamination. However, there are some materials that are kinetics limited, especially at low temperature.

Gaseous impurity resistance is a very important property, especially when the application is "open-ended" and uses new H_2 for each Hydriding/Dehydriding cycle, that H_2 often being impure. Depending on the alloy impurity combination, there can be several types of damage, such as Poisoning, Retardation, Reaction and Innocuous.

Usually, damages from poisoning and retardation are usually recoverable, but reaction damage is usually not. Cyclic stability is important and widely variable from alloy to alloy. Alloys and intermetallic compounds are usually metastable relative to disproportionation, the tendency to break up metallurgically (phase transformation) to form stable, not easily reversed hydrides. Even if very pure H_2 is used, disproportionation can occur with a resultant loss of reversible capacity. Safety usually centres on ignition, the tendency for a hydride powder to burn when suddenly exposed to air, e.g. an accidental tank rupture. But the term can also include toxicity resulting from accidental ingestion or inhalation. Alloy cost is influenced by several factors, including raw materials cost, melting and annealing costs, metallurgical complexities, profit and the degree of PCT precision needed for the particular application. It is important to point-out that the above factors in addition to raw materials cost can easily raise true alloy cost by more than 100%.

In addition to the AB_5, AB_2, AB and A_2B intermetallic compounds discussed above, several other families of intermetallics have been shown capable of reversible hydriding/dehydriding reactions. Examples include AB_3, A_2B_7, A_6B_{23}, A_2B_{17}, A_3B and others. Most structures involve long-period AB_5 and AB_2 stacking sequences and are thus crystallographically related to these two classic families. Although none of these have attained commercial levels of interest, at least the AB_3 and A_2B_7 phases do have PCT properties, which are in the range of our interest. Most either have narrow plateaux with long sloping upper legs (e.g. $GdFe_3$) or multiple plateaux (e.g. $NdCo_3$ or Pr_2Ni_7). La_2Mg_{17} was once reported to have 6 wt% H-capacity, recoverable at room temperature, but that claim has never been independently confirmed.

The solution of intermetallic compounds, which represent the hydrogen storage absorbed on interstitial sites in a host metal, will be further discussed. Research on intermetallic compounds for hydrogen storage has been commenced more than 30 years ago and it has always been on the foreground. The first discovery of hydrogen absorption by ZrNi [31], $LaNi_5$ [32, 30] and TiFe [33] opened new possibilities for industrial developments. However, for on-board storage, they remained at the stage of prototypes due to their weight penalty and low hydrogen storage capacity, since they don't meet the criteria determined by energy commissions around the world [34–39].

The host materials in the intermetallic compounds are ordered stoichiometric compounds, typically formed from two metallic components, A and B. As it is illustrated in Figure 1.9, intermetallic compounds are built by alloying a metal which easily forms stable hydrides (A) and another element which does not form stable hydrides (B). The components A and B can generally be fully or partially substituted by other elements of relatively similar size or chemistry. The intermetallics thus formed could then be grouped according to their stoichiometry such as AB_5, AB_2, AB and others. Hydrogen absorbing intermetallics form a number of different groups as shown in Table 1.2.

In general, metals are crystallised in one of the following structures in near ambient conditions (Figure 1.10):

- Body Centred Cubic (BCC)
- Face Centred Cubic (FCC)
- Hexagonal Closed Packed (HCP)

More specifically, many intermetallic compounds of composition AB mainly crystallise in the BCC structure, while AB_2 crystallise in one of three closely related structures (Laves Phases), which are:

- The hexagonal C14 structure, typified by the phase $MgZn_2$
- The cubic C15 structure, typified by the phase $MgCu_2$
- The hexagonal C36 structure, typified by the phase $MgNi_2$

These components, the "Laves phases," involve the association of large A atoms and small B atoms. Since the ratio of the atomic diameters of the components

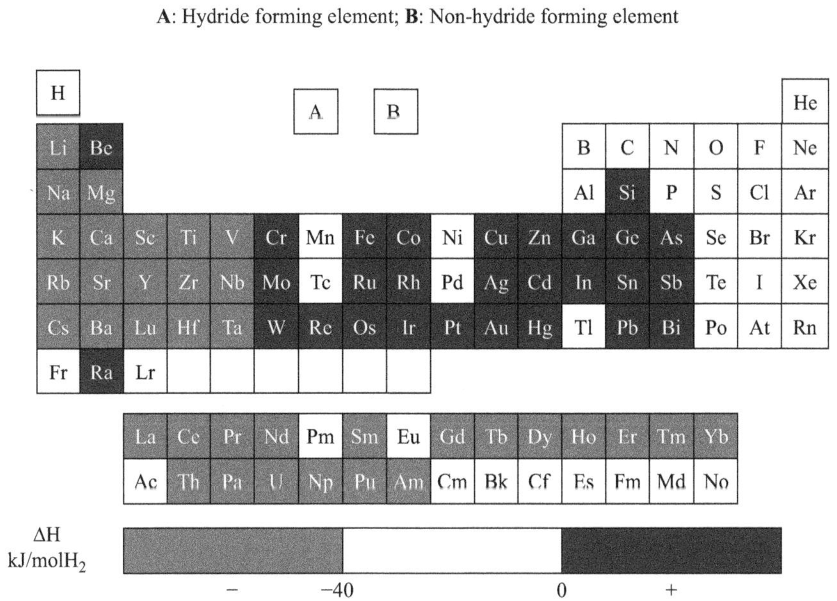

Figure 1.9 Hydride and non-hydride forming elements in the periodic system of elements [40]

Table 1.2 Examples of intermetallic hydrides [20, 40]

Composition	A	B	Compounds
AB_5	Ca, La, rare earth	Ni, Cu, Co, Pt, Fe	$CaNi_5$, $LaNi_5$, $CeNi_5$, $LaCu_5$, $LaPt_5$, $LaFe_5$
AB_2	Zr, Ti, Y, La	V, Cr, Mn, Fe, Ni	$LaNi_2$, YNi_2,YMn_2, $ZrCr_2$, $ZrMn_2$, ZrV_2, $TiMn_2$
AB_3	La, Y, Mg	Ni, Co	$LaCo_3$, YNi_3, $LaMg_2Ni_9$
A_2B_7	Y, Th	Ni, Fe	Y_2Ni_7, Th_2Fe_7
A_6B_{23}	Y	Fe	Y_6Fe_{23}
AB	Ti, Zr	Ni, Fe	TiNi, TiFe, ZrNi
A_2B	Mg, Zr	Ni, Fe, Co	Mg_2Ni, Mg_2Co, Zr_2Fe

d_A/d_B is approximately 1.2, the phases are regarded as controlled essentially by atomic size relationships. This interpretation is supported by the observation that the A and B components may be taken from any group of the Periodic Table, while the same element may form either A or the B component in different compounds [42].

During hydrogenation, hydrogen atoms will occupy specific interstitial sites. The interstitial sites in the three major crystal structures of hydrides are shown in Figure 1.10. Only octahedral (O) and tetrahedral (T) sites are shown because they are the only ones occupied by hydrogen atoms. Nevertheless, some distinction should be

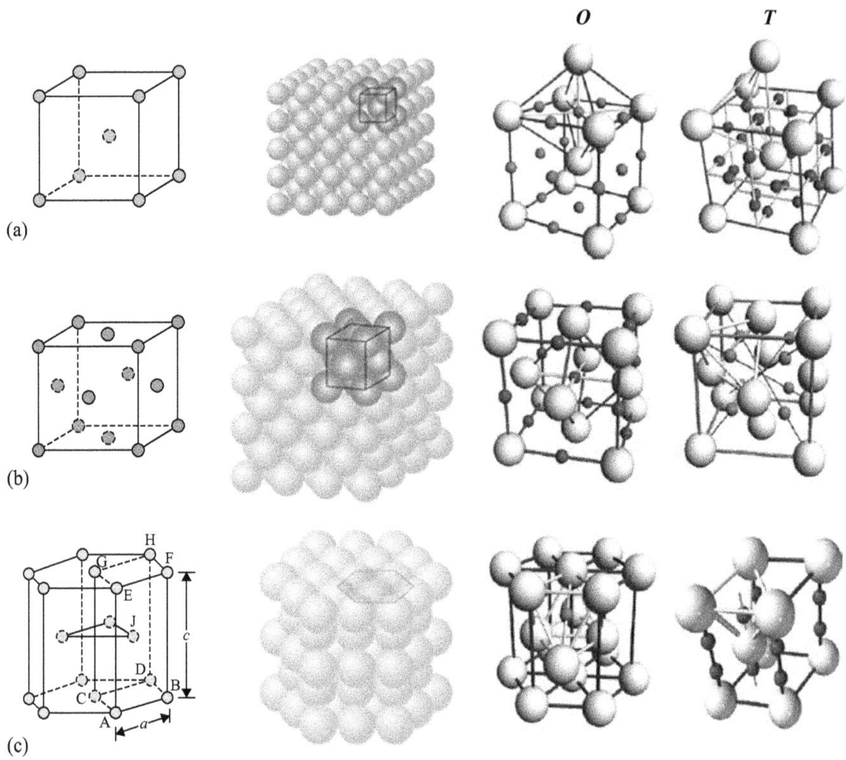

Figure 1.10 Schematic representation for the (a) Body Centred Cubic (BCC), (b) Face Centred Cubic (FCC) and (c) Hexagonal Closed Packed (HCP) structure, and their interstitial octahedral (O) and tetrahedral (T) sites [10, 41]

made between the different crystal structures. In the FCC lattice, the T and O sites are enclosed in regular tetrahedral and octahedral, respectively, formed by metal atoms. In the HCP lattice, the tetrahedral or octahedral sites become distorted as the ratio of lattice parameters c/a deviates from the ideal value of 1.633 [10].

An important feature of the AB_2 alloys (and to a lower extent, the AB alloys) is that their operating characteristics (equilibrium pressure and temperature) can be adjusted through solute substitution for both A- and B-site metals. The equilibrium pressures at a selected temperature can often be changed by more than two orders of magnitude. In general, the larger the lattice parameters of the host alloy, the lower the pressure. Deviations in the alloy composition (i.e. changes in the B/A ratios) also lead to significant changes in the PCI characteristics, which can be advantageous in many situations. Unfortunately, both of these alloy variations usually result in a smaller hydrogen storage capacity than pure stoichiometric binary AB_2 intermetallics.

1.3.3.1 AB-type TiFe intermetallic compounds

Among hydrogen storage materials, TiFe has the advantages of a large absorption capacity (1.8–1.9 wt%), good kinetics for sorption and desorption of hydrogen after activation, suitable equilibrium pressures (5–10 bar) and low cost. However, TiFe has the great disadvantage of not being easily activated at room temperature [33], [43–47]. But despite the advantages of TiFe, its weak points must be overwhelmed in order for this compound to find a practical application. As it is previously mentioned, a common method to modify the hydrogen sorption properties of a hydride is the insertion of substituting elements. For this compound, V, Mn or/and other transition metals are widely used to partially substitute Fe. For example, when a small amount of Mn is substituted for Fe in the TiFe parent compound, activation is promoted and the resistance to contamination by impure gases is increased. TiFe intermetallic compound counts down a history of about 40 years since its discovery as a potential hydrogen storage material and has been thoroughly studied ever since in an effort to improve its properties [48–52].

1.3.3.2 AB_2 Laves Phases Zr-based intermetallic compounds

Laves phases form the largest group of intermetallics and thus have a wide range of properties. Their stability depends on various factors such as: geometry, packing density, valence electron concentration or the difference in electronegativity. Laves phases have been recognised to be attractive hydrogen storage materials, particularly the Zr-based alloys. They have relatively good hydrogen storage capacity and kinetics, long cycling life and low cost but they are usually too stable at room temperature and are sensitive to gas impurities. Shaltiel *et al.* [54] first studied the hydrogenation characteristics of various Zr-based Laves Phase compounds such as ZrV_2, $ZrCr_2$, $ZrMn_2$, $ZrFe_2$ and $ZrCo_2$. Most of these alloys could absorb large quantities of hydrogen but the hydrides formed were too stable to be of practical significance. Therefore many studies performed so far have paid attention to increasing the equilibrium pressure of Zr-based Laves Phase compounds, without modifying the absorption capacity, by partial substitution of A or B element by other elements. Substitution usually involves transition elements, creating multi-component systems ($(Zr_{1-x}X_x(Fe, Mn, Cr)_{2-y}Y_y$, where X = Ti, Y, Hf, Sc, Nb and Y = V, Mo, Mn, Cr, Fe, Co). However, in many cases Ti and Fe are being mostly used for the substitution of A and B elements, respectively [10], [53–59].

1.4 Hydrogen compressors

Hydrogen compression is achieved by a mechanical device that increases the pressure of the gas by reducing its volume. However, the low molar weight of hydrogen requires the use of a volumetric compressor rather than a centrifugal one in order to gain efficiency. Moreover, the energy used to compress gas does not only produce a pressure increase, but also generates heat. Hydrogen compressors are categorised in mechanical, non-mechanical hydrogen compressors. Their properties and characteristics will be further described.

1.4.1 Mechanical compressors

The piston compressor is an electro-hydraulically driven, non-lubricated, liquid-cooled, single-stage unit, including an electric motor, hydraulic oil tank, high pressure gas intensifier and intensifier shifting mechanism. As it can be seen in Figure 1.11, representing a single-stage unit, the intensifier contains a hydraulic drive cylinder in the centre that is coupled by tie rods with two single-stage gas cylinders on each side. The fluid power drive provides the intensifier with high pressure hydraulic oil. During operation, the gas fills the cylinder and then, the force of the hydraulic pressure acts on the hydraulic piston compressing the gas in the cylinder. Once compression is completed, the four-way valve redirects the hydraulic fluid and the piston assembly moves in the opposite direction.

In the case of a piston-metal diaphragm hydrogen compressor, the gas from the piston and related components is isolated by a set of metal diaphragms as shown in Figure 1.12. The piston moves a column of hydraulic fluid, which in turn moves the diaphragm set and displaces the gas to be compressed. As a consequence, the process is more isentropic than adiabatic, thus allowing to achieve a higher compression ratio [11].

1.4.2 Non-mechanical compressors

These types of compressors have several advantages over mechanical hydrogen compressors, including smaller size, lower capital, operating and maintenance costs and the absence of moving parts, which eliminates problems related to wear, noise and intensity of energy usage. Since hydrogen is completely separated from the hydraulic fluid, high purity hydrogen can be supplied [11].

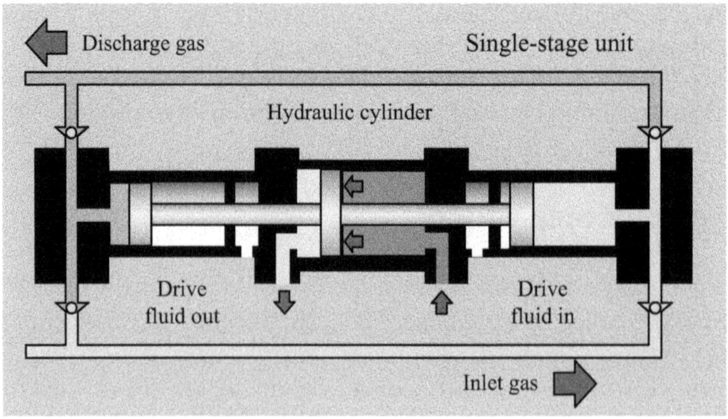

Figure 1.11 Schematic diagram of a single-stage compression unit [11]

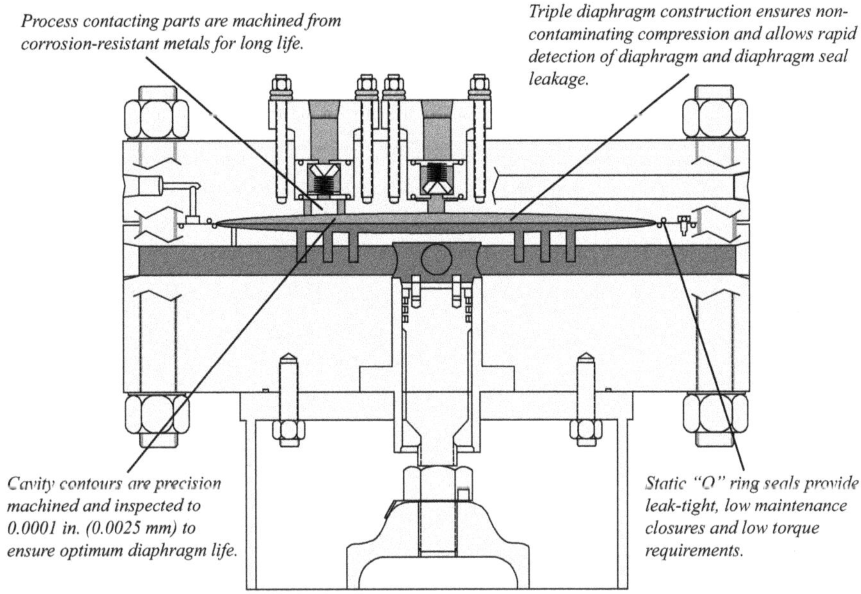

Figure 1.12 Schematic diagram of a metal diaphragm hydrogen compressor [47]

As mentioned earlier, this category of compressors is represented by two types of devices, the electrochemical and the metal hydride hydrogen compressor [60, 61]. Within the aim of this study, the electrochemical compressor will be briefly presented while the metal hydride hydrogen compressors will be thoroughly discussed.

The **electrochemical hydrogen compressor** is used when small quantities of hydrogen have to be delivered in high pressure, because it is more efficient than the mechanical compressor in this regime [42]. The working principle is based on an electrochemical cell, composed of an anode, a membrane electrode assembly and a cathode, as it schematically represented in Figure 1.13. When a potential difference is applied, the hydrogen at a pressure Pa is oxidised to H+ at the anode. These ions are transported through the membrane to the cathode, where they are reduced to hydrogen at a pressure Pc > Pa, if the cathode compartment is hermetically sealed. As long as hydrogen and power are supplied, this electrochemical reaction continues to compress hydrogen. A multi-stage electrochemical hydrogen compressor incorporates a series of membrane electrode assemblies. It should be noted that the process is selective for hydrogen, as the inert gas components cannot pass the membrane [11].

On the other hand, **metal hydride hydrogen compressors** are efficient and reliable thermally powered systems that use the properties of reversible metal hydride alloys to compress hydrogen without contamination. The most important component of this thermal engine is the metal hydride. A wide range of different operation characteristics can be obtained by either varying the type of the hydride or just modifying the formula of a reference alloy. The operating principle of the metal hydride hydrogen compressor is based on heat and mass transfer (hydrogen)

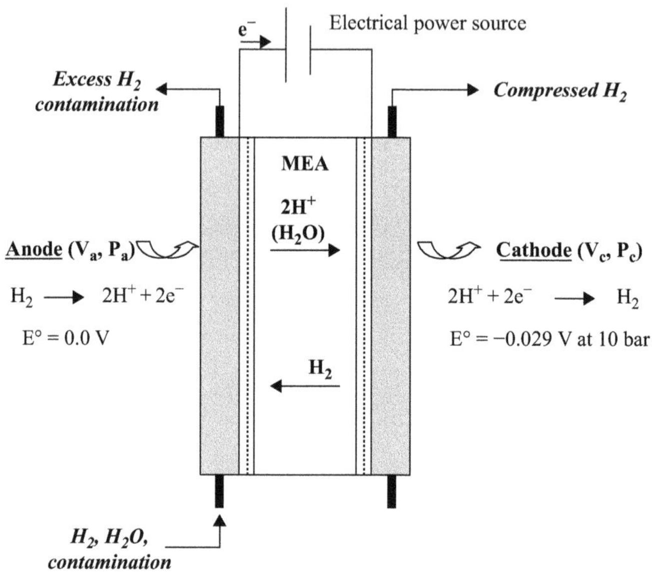

Figure 1.13 Principle of the electrochemical hydrogen compressor [62]

in the reaction bed during absorption and desorption process [11]. The advantages of metal hydride compression include simplicity in design and operation, absence of moving parts, compactness, safety and reliability, and on the most important, the possibility to consume waste industrial heat instead of electricity [63].

The first studies on metal hydride hydrogen compressors have been reported already from 1984, and since then numerous R&D activities on the development of such a device have been published in several forms, as papers, conference presentations, reports and patents [63–83].

Since this type of compressor is based on metal hydrides, as a consequent, it follows the rules of function of hydrides. That means that they have in common all the advantages and drawbacks. But more specifically, metal hydride compressors are simple and efficient pressure/temperature swing absorption-desorption systems. This allows not only controlling pressure by varying the temperature, but it also opens a new horizon for hydrogen separation and purification [63].

The selection of an adapted metal hydride alloy is important to reach the desired performance of the compression cycle. The thermodynamic (ΔH and ΔS), thermal (specific heat C_p) and kinetic properties of the selected material should be characterised in detail in order to optimise compressor operation.

A single-stage thermal compressor is composed of a module filled with a reversible metal hydride alloy. Figure 1.14 displays the operation of the hydride compressor, which comprises four processes [11]:

Point **A**: Hydrogen is absorbed into the alloy bed at low temperature (T_c) and at low pressure (P_s).

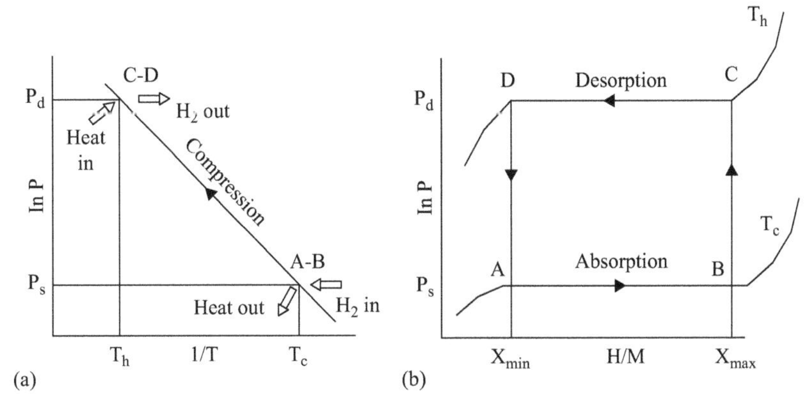

Figure 1.14 Principle of operation of a single stage metal hydride hydrogen compressor [75]

Point **B**: The module is subsequently heated from (T_c) to a higher temperature (T_h) with compression taking place at the same time.
Point **C**: Compressed hydrogen is released at high temperature (T_h) and high pressure (P_d).
Point **D**: The module is cooled down (from T_h to T_c).

In a multi-stage metal hydride compressor, hydrogen is successively absorbed and desorbed into and out of several hydride beds. The hydrides operate between two temperature levels. The warm side provides the heat for desorption of hydrogen, and the cold side is the heat sink to dissipate the heat generated when hydrogen is absorbed. The hydride alloys are selected in such a way that the lower pressure hydride desorbs hydrogen at high temperature into the higher pressure hydride which absorbs hydrogen at low temperature (Figure 1.15) [84].

But the selection of the alloys should take into account more factors, mainly concerning the operation conditions of the device. Thus, before the selection it is essential to set the boundary conditions, such as:

- **Cold side temperature** (or temperature of cooling). The value of this parameter is usually around ambient temperature or less, depending on the procedure and the resources available for the cooling.
- **Hot side temperature** (or temperature of heating). Since water is mainly used for heating and cooling as well, the value of this parameter is not usually exceeding 90 °C, but if another way of cooling and heating is installed, this temperature can be as high as the metal hydride thermodynamic properties allow. On the other hand, there is an excess of waste heat from other processes that is available, but it hardly surpasses 70–80 °C (waste heat of an electrolyser and fuel cell/solar collector).
- **Input pressure**. This parameter is independent only if the metal hydride hydrogen compressor is not coupled with another device. But in order to increase the overall efficiency of the device, the aim is to couple such a

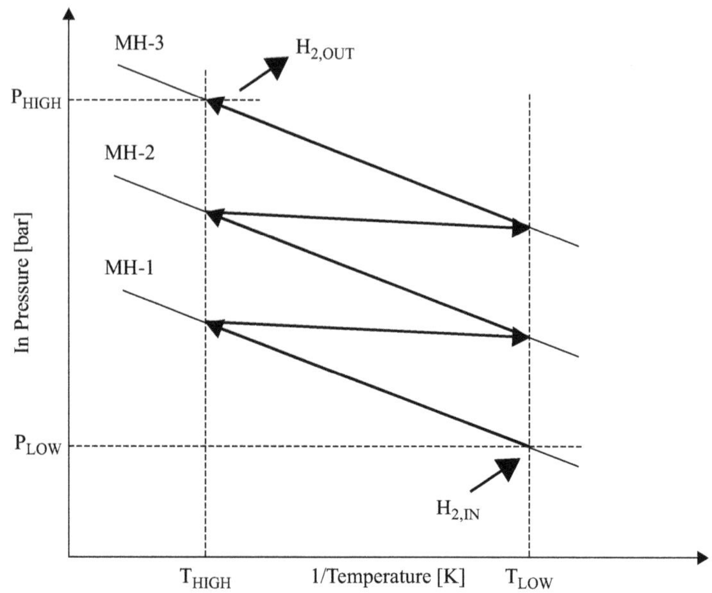

Figure 1.15 Principle of a three-stage metal hydride hydrogen compressor [71]

compressor with an electrolyser; the input pressure must be in the range of the output pressure of the electrolyser, i.e. 10–30 bar.

- **Output pressure**. The value of this parameter is dependent of the technical specifications of the device, which includes the hot side temperature and the capacity of the hydride beds to withstand the temperature and the pressure demanded.

When the discussion comes to the overall hydrogen compression by metal hydrides unit, more factors must be taken into account. The major factor affecting hydrogen compression efficiency is the plateau slope of the materials selected. In multi-component intermetallic compounds the sloping plateaux are originated form compositional fluctuations due to the presence of impurities or because of fluctuations of the stoichiometry ($AB_{n\pm x}$) within the homogeneity region [85]. In addition to operating pressure-temperature ranges, an important parameter of the material for hydrogen compression is the process productivity. Another important factor for the hydrogen compression performance of the real metal hydride systems is hysteresis, as the values of plateau pressures for hydrogen absorption/hydrogenation are higher than the ones for hydrogen desorption/dehydrogenation. Hysteresis is caused by various reasons; the most frequent one is the stresses which appear in the course of growth of hydride nuclei inside the matrix of the metal alloy having lower molar volume [63].

Many groups have been focused in hydrogen compression metal hydrides materials systems as it is recognised as one of the potential efficient ways of hydrogen storage technology in mobile applications [86–90].

Acknowledgements

I would like to thank my ex-PhD students, Dr Alexandra Ioannidou, Dr Evangelos Gkanas and Dr Evangelos Koultoukis, for their collaboration over the last decade in metal hydride storage and compression experimental and theoretical research work.

References

[1] Kelly NA, Gibson TL, Cai M, Spearot JA, Ouwerkerk DB, Development of a renewable hydrogen economy: optimization of existing technologies. *International Journal of Hydrogen Energy* 2010;35:892–899.
[2] Crabtree GW, Dresselhaus MS, Buchanan MV, The hydrogen economy. *Physics Today* 2004;57:39–44.
[3] Mac E, Gray A, Hydrogen storage – status and prospects. *Advances in Applied Ceramics* 2007;106:25–28.
[4] Wolf D, Budt M, LTA-CAES – a low-temperature approach to adiabatic compressed air energy storage. *Applied Energy* 2014;125:158–164.
[5] Sebastián R, Peña Alzola R, Flywheel energy storage systems: review and simulation for an isolated wind power system. *Renewable and Sustainable Energy Reviews* 2012;16:6803–6813.
[6] Sioshansi F (ed.). *Generating Electricity in a Carbon-Constrained World.* Elsevier Inc, Cambridge, Massachusetts; 2009.
[7] Melhem Z (ed.). *Electricity transmission, distribution and storage systems.* Woodhead Publishing Limited, Sawston, Cambridge; 2013.
[8] Jain IP, Hydrogen, the fuel for the 21st century. *International Journal of Hydrogen Energy* 2009;34:7368–7378.
[9] Schlapbach L, Zuttel A, Hydrogen-storage materials for mobile applications. *Nature* 2001;414:353–358.
[10] Hirscher M (ed.). *Handbook of hydrogen storage: new materials for future energy storage.* Weinheim: Wiley-VCH Verlag GmbH & Co. KgaA; 2010.
[11] Léon A (ed.). *Hydrogen technology.* Berlin Heidelberg: Springer-Verlag; 2008.
[12] Linde AG, Germany, http://www.linde.com/en/index.html.
[13] McPhy Energy SA, France, http://www. mcphy.com/en/.
[14] Bossel U, Eliasson B, The future of the hydrogen economy: bright or bleak? *Proceedings, THE FUEL CELL WORLD*, Lucerne/Switzerland; July 2002.
[15] Bossel U, The physics of the hydrogen economy. *European Fuel Cell News* 2003;10(2).
[16] Cleveland CJ, Chronologies, top ten lists, and world clouds. *Handbook of Energy* 2014;2:309–322.
[17] Dunn S, History of hydrogen. *Encyclopaedia of Energy* 2004;3:241–252.
[18] Lide DR (ed.). *CRC handbook of chemistry and physics.* Boca Raton, FL: CRC Press; 2005.
[19] *Hydrogen properties, course manual on hydrogen fuel cell engines and related technologies, Module 1*, College of the Desert, Palm Desert, CA, USA; 2001.

[20] Züttel A, Materials for hydrogen storage, *Materials Today*; 2003.

[21] Tzimas E, Filiou C, Peteves SD, Veyret JB, *Hydrogen storage: state-of-the-art and future perspective*. Mission of the Institute for Energy, European Communities; 2003.

[22] Bossel U, Eliasson B, *Energy and the hydrogen economy, US DOE, EERE*, http://www.afdc.energy.gov/pdfs/hyd_economy_bossel_eliasson.pdf.

[23] Durbin DJ, Malardier-Jugroot C, Review of hydrogen storage techniques for on board vehicle applications. *International Journal of Hydrogen Energy* 2013;38:15595–14617.

[24] Sakintura B, Lamari-Darkrim F, Hirscher M, Metal hydride materials for solid hydrogen storage: a review. *International Journal of Hydrogen Energy* 2007;32:1121–1140.

[25] Von Helmolt R, Eberle U, Fuel cell vehicles: status 2007. *Journal of Power Sources* 2007;165:833–843.

[26] *Encyclopédie des gaz, Air Liquide, Division scientifique*. Amsterdam: Elsevier; 1976.

[27] Gstrein G, Klell M, Properties of hydrogen. Institute for Internal Combustion Engines and Thermodynamics, Graz University of Technology; 2004.

[28] Weast RC, Astle MJ, Beyer WH, *CRC handbook of chemistry and physics*. Boca Raton, FL: CRC Press; 1983.

[29] Van Midden HJP, Prodan A, Zupanic E, Zitko R, Makridis SS, Stubos AK, Structural and electronic properties of the hydrogenated $ZrCr_2$ Laves phases. *Journal of Physical Chemistry* 2010;114:4221–4227.

[30] Schlapbach L, Anderson I, Burger JP, Hydrogen in metals. *Electronic and magnetic properties of metals and ceramics Part II*, vol 3B. Weinheim: Wiley VCH; 1994.

[31] Lobowitz GG, Hayes HF, Gibb TRP, The system zirconium–nickel and hydrogen. *Journal of Physical Chemistry* 1958;62:76–79.

[32] Vucht JV, Kuijpers FA, Bruning H, Reversible room-temperature absorption of large quantities of hydrogen by intermetallic compounds. *Philips Research Reports* 1970;25:133–140.

[33] Reilly JJ, Wiswall RH, Formation and properties of iron titanium hydride. *Inorganic Chemistry* 1974;13:218–222.

[34] Reilly JJ, Sandrock GD, Hydrogen storage in metal hydrides. *Scientific American* 1980;13:118–119.

[35] Alefeld G, Völkl J (eds). *Topics in applied physics vol.28: hydrogen in metals I. Basic properties*. Berlin: Springer-Verlag; 1978.

[36] Alefeld G, Völkl J (eds). *Topics in applied physics vol.29: hydrogen in metals II. Application-oriented properties*. Berlin: Springer-Verlag; 1978.

[37] Schlapbach L (ed). *Topics in applied physics vol.63: hydrogen in intermetallic compounds I. Electronic, thermodynamic and crystallographic properties, preparation*. Berlin: Springer-Verlag; 1988.

[38] Schlapbach L (ed). *Topics in applied physics vol.67: hydrogen in intermetallic compounds II. Surface and dynamic properties, applications*. Berlin: Springer-Verlag; 1992.

[39] Wipf H (ed). *Topics in applied physics vol.73: hydrogen in metals III. Properties and applications*. Berlin: Springer-Verlag; 1997.
[40] Moreno-Pirajan JC (ed.) *Thermodynamics – Interaction Studies – Solids, Liquids and Gases*. InTech, 2011.
[41] Callister WD, *Materials science and engineering: an introduction*. New York: John Wiley & Sons, Inc; 2007.
[42] Berry RL, Raynor GV, The crystal chemistry of the Laves Phases. *Acta Crystallographica* 1953;6:178–186.
[43] Thompson P, Reilly JJ, Hastings JM, The application of the Rietveld method to a highly strained material with microtwins: $TiFeD_{1.9}$. *Journal of Applied Crystallography* 1989;22:256–260.
[44] Thompson P, Reidinger F, Reilly JJ, Corliss LM, Hastings JM, Neutron diffraction study of α iron titanium deuteride. *Journal of Physics F* 1980;10: L57–L59.
[45] Thompson P, Pick MA, Reidinger F, Corliss LM, Hastings JM, Reilly JJ, Neutron diffraction study of β iron titanium deuteride. *Journal of Physics F* 1978;8(4):L75–L80.
[46] Thompson P, Reilly JJ, Reidinger F, Hastings JM, Corliss LM, Neutron diffraction study of γ iron titanium deuteride. *Journal of Physics F* 1979; 9(4):L61–L66.
[47] Fischer P, Schefer J, Yvon K, Schlapbach L, Riesterer T, Orthorhombic structure of γ-$TiFeD_2$. *Journal of Less Common Metals* 1987;129:39–45.
[48] Miller HI, Murray J, Laury E, Reinhardt J, Goudy AJ, The hydriding and dehydriding kinetics of FeTi and $Fe_{0.9}TiMn_{0.1}$. *Journal of Alloys and Compounds* 1995;231:670–674.
[49] Kinaci A, Aydinol MK, Ab initio investigation of FeTi-H system. *International Journal of Hydrogen Energy* 2007;32:2466–2474.
[50] Yukawa H, Matsumura T, Morigana M, Chemical bond state and hydride stability of hydrogen storage alloys. *Journal of Alloys and Compounds* 1999;293:227–230.
[51] Williams M, Lototsky MV, Davids MW, Linkov V, Yatrys VA, Solberg JK, Chemical surface modification for the improvement of the hydrogenation kinetics and poisoning resistance of TiFe. *Journal of Alloys and Compounds* 2011;509S:770–774.
[52] Endo N, Saitoh H, Machida A, Katayama Y, Formation of BCC TiFe hydride under high hydrogen pressure. *International Journal of Hydrogen Energy* 2013;38:6726–6729.
[53] Shaltiel D, Hydride properties of AB_2 Laves Phase compounds. *Journal of Less Common Metals* 1978;62:407–416.
[54] Shaltiel D, Jacob I, Davidov D, Hydrogen absorption and desorption properties of AB_2 Laves-Phase pseudobinary compounds. *Journal of Less Common Metals* 1977;53:117–131.
[55] Zhang TB, Wang XF, Hu R, *et al.*, Hydrogen absorption properties of Zr $(V_{1-x}Fe_x)_2$ intermetallic compounds. *International Journal of Hydrogen Energy* 2012;37:2328–2335.

[56] Yadav TP, Shahi RR, Srivastava ON. Synthesis, characterization and hydrogen storage behaviour of AB_2 ($ZrFe_2$, $Zr(Fe_{0.75}V_{0.25})_2$, $Zr(Fe_{0.5}V_{0.5})_2$ type materials. *International Journal of Hydrogen Energy* 2012;37: 3689–3696.

[57] Hangström MT, Klyamkin SN, Mescheryakova EV, Lund PD, High-pressure AB_2 metal hydrides with low hysteresis. *Journal of Materials Science* 2000;35:127–131.

[58] Zotov TA, Sivov RB, Movlaev EA, Mitrokhin SV, Verbetsky VN, IMC hydrides with high hydrogen dissociation pressure. *Journal of Alloys and Compounds* 2011;509S:839–843.

[59] Yoshida M, Akiba E, Hydrogen absorbing-desorbing properties and crystal structure of the Zr-Ti-Ni-Mn-V AB_2 Laves phase alloys. *Journal of Alloys and Compounds* 1995;224:121–126.

[60] Pressure Products Industries, Milton Roy, USA, http://www.pressure productsindustries.com/.

[61] Rohland B, Eberle K, Ströbel R, Scholta J, Garche J, Electrochemical hydrogen compressor. *Electrochimica Acta* 1998;43:3841–3846.

[62] Ströbel R, Oszcipok M, Fasil M, Rohland B, Jörissen L, Garche J, The compression of hydrogen in an electrochemical cell based on a PE fuel cell design. *Journal of Power Sources* 2002;105:208–215.

[63] Lototskyy MV, Yartys VA, Pollet BG, Bowman Jr. RC, Metal hydride hydrogen compressors: a review. *International Journal of Hydrogen Energy* 2014;11:5818–5851.

[64] Lynch JF, Maeland AJ, Libowitz GG, Hydrogen compression by metal hydrides. In: Veziroglu TN, Taylor IB, editors. *Hydrogen energy progress V. Proceedings of the 5th world hydrogen energy conference*. vol.1. Oxford: Pregamon Press; 1984.

[65] Dantzer P, Orgaz E, Thermodynamics of the hydride chemical heat pump: model I. *Journal of Chemical Physics* 1986;85:2961–2973.

[66] Dantzer P, Orgaz E, Thermodynamics of the hydride chemical heat pump: II. How to select a pair of alloys. *International Journal of Hydrogen Energy* 1986;11:797–806.

[67] Orgaz E, Dantzer P, Thermodynamics of the hydride chemical heat pump: III. Considerations for multistage operation. *Journal of Less-Common Metals* 1987;131:385–398.

[68] Solovey VV, Metal hydride thermal power installations. In: Veziroglu TN, Protsenko AN, editors. *Hydrogen energy progress VII. Proceedings of the 7th world hydrogen energy conference*. vol. 2. Oxford: Pergamon Press; 1988.

[69] Sandrock G, Applications of hydrides. In: Yurum Y, editor. *Hydrogen energy system. Production and utilization of hydrogen and future aspects.* Kluwer Acad Publication. NATO ASI Ser. Ser E, 295; 1995.

[70] Dantzer P, Metal-hydride technology: a critical review. In: Wipf H, editor. *Hydrogen in metals III. Properties and applications.* Berlin-Heidelberg: Springer-Verlag; 1997.

[71] Vanhanen JP, Hagström MT, Lund PD, Combined hydrogen compressing and heat transforming through metal hydrides. *International Journal of Hydrogen Energy* 1999;24:330–337.
[72] Bowman Jr RC, Kichl B, Marquardt E, Closed cycle Joule Thomson cryo coolers. In: Donabedian M, editor. *Spacecraft thermal control handbook. Cryogenics*. vol. 2. El Segundo, CA, USA: The Aerospace Press; 2003.
[73] Bowman Jr RC, Development of metal hydride beds for sorption cryocoolers in space applications. *Journal of Alloys Compounds* 2003;356–357:789–793.
[74] Dehouche Z, Grimard N, Laurencelle F, Goyette J, Bose TK, Hydride alloys properties investigations for hydrogen sorption compressor. *Journal of Alloys and Compounds* 2005;399:224–236.
[75] Muthukumara P, Maiya MP, Murthy SS, Experiments on a metal hydride based hydrogen compressor. *International Journal of Hydrogen Energy* 2005;30:879–892.
[76] Laurencelle F, Dehouche Z, Goyette J, Bose TK, Integrated electrolyser-metal hydride compression system. *International Journal of Hydrogen Energy* 2006;31:762–768.
[77] Wang X, Chen R, Zhang Y, Chen C, Wang Q, Hydrogen storage alloys for high-pressure supra-pure hydrogen compressor. *Journal of Alloys and Compounds* 2006;420:322–325.
[78] Muthukumara P, Maiya MP, Murthy SS, Performance tests on a thermally operated hydrogen compressor. *International Journal of Hydrogen Energy* 2008;33:463–469.
[79] Laurencelle F, Dehouche Z, Morin F, Goyette J, Experimental study on a metal hydride based hydrogen compressor. *Journal of Alloys and Compounds* 2009;475:810–816.
[80] Popeneciu G, Almasan V, Coldea I, Lupu D, Misan I, Ardelean O, Investigation on a three-stage hydrogen thermal compressor based on metal hydrides. Processes in isotopes and molecules. *Journal of Physics: Conference Series* 2009;182:012053.
[81] Hui L, Xinhua W, Zhaohui D, Lou X, Changpin C, A study on 70MPa metal hydride hydrogen compressor. *Journal of Alloys and Compounds* 2010;502:503–507.
[82] Hopkins RR, Kim KJ, Hydrogen compression characteristics of a dual stage thermal compressor system utilizing $LaNi_5$ and $Ca_{0.6}Mm_{0.4}Ni_5$ as the working metal hydrides. *International Journal of Hydrogen Energy* 2010;35:5693–5702.
[83] Muthukumar P, Patel KS, Sachan P, Singhal N, Computational study on metal hydride based three-stage hydrogen compressor. *International Journal of Hydrogen Energy* 2012;37:3797–3806.
[84] Golben PM, Multi-stage hydride-hydrogen compressor. *Proceedings of the 18th Intersociety Energy Conversion Engineering Conference*; 1983.
[85] Shilov AL, Efremenko NE, Effect of sloping pressure "plateau" in two-phase regions of hydride systems. *Russian Journal of Physical Chemistry* 1986;60:3024–3028.

[86] Koultoukis ED, Makridis SS, Pavlidou E, de Rango P, Stubos AK, Investigation of $ZrFe_2$-type materials for metal hydride hydrogen compressor systems by substituting Fe with Cr or V. *International Journal of Hydrogen Energy* 2014;39(36):21380–21385.

[87] Koultoukis ED, Gkanas EI, Makridis SS, Christodoulou CN, Fruchard D, Stubos AK, High temperature activated AB_2 nanopowders for metal hydride hydrogen compression. *International Journal of Energy Research* 2014;38 (4):477–486. Doi:10.1002/er.3147.

[88] Gkanas EI, Makridis SS, Stubos AK, Modeling and simulation for absorption-desorption cyclic process on a three-stage metal hydride hydrogen compressor. *Computer Aided Chemical Engineering* 2013;32:379–384.

[89] Xiao J, Bénard P, Chahine R, Charge-discharge cycle thermodynamics for compression hydrogen storage system. *International Journal of Hydrogen Energy* 2016. doi: doi:10.1016/j.ijhydene.2015.12.136.

[90] Lototskyy MV, Yartys VA, Pollet BG, Bowman Jr. RC, Metal hydride hydrogen compressors: a review. *International Journal of Hydrogen Energy* 2014;39(11):5818–5851.

Chapter 2

Power-to-Gas for energy system integration and storage

Sean B. Walker[1], Ushnik Mukherjee[1], Michael Fowler[1], Daniel van Lanen[1] and Azadeh Maroufmashat[1]

2.1 Summary of concept

In Ontario, utilities are working to manage the surplus base load power produced by low-emissions nuclear power generation and the increasing penetration of intermittent renewable sources by developing a Smart Grid where energy can be stored and distributed as needed. Transforming electricity into hydrogen gas offers a potential system-wide, distributed energy storage system, which can be used to handle surplus power, off-peak periods or intermittent renewable power generation. The conversion of electrical energy to gas energy through a concept called 'Power-to-Gas (PtG)', is implemented through the use of electrolyzers and allows simultaneous grid stabilization, seasonal storage of bulk power, geographic transmission of energy and dispatchable regeneration of distributed renewable energy [1].

A key advantage of the development of PtG is the ability to move energy between the electrical and natural gas (NG) systems. This bi-directional energy flow adds flexibility that is used to facilitate the development of Smart Grid integrated energy system. The conversion of renewable and surplus power to hydrogen through PtG optimizes both the existing NG and electricity networks to limit wasted energy and costly electricity exports while providing an opportunity to integrate more renewable energy into the electrical grid. In addition to the economic and environmental benefits of PtG, optimizing the existing infrastructure by deferring the infrastructure costs associated with the construction of new electricity transmission infrastructure will increase public acceptance.

As shown in Figure 2.1, PtG is an energy storage and transmission solution which utilizes multiple energy sources to produce hydrogen gas that can be injected into the NG infrastructure, used on-site or stored in tanks or reservoirs. The PtG system may also be used to move energy from one location to another, where it can then be used to generate electricity or heat on demand, more effectively. The only new infrastructure required by the system is the aforementioned electrolyzer.

[1]Department of Chemical Engineering, University of Waterloo, Waterloo, Ontario, Canada

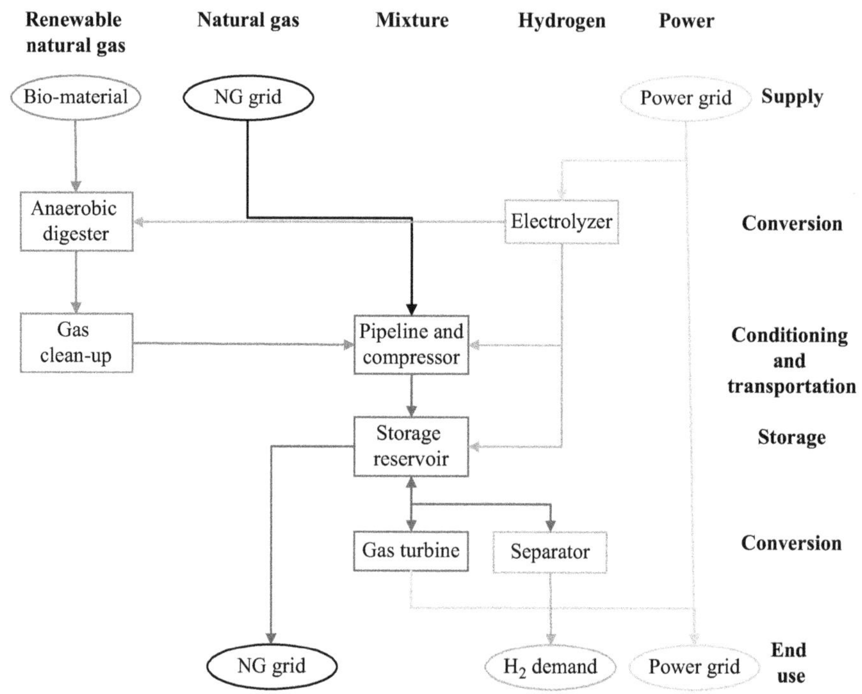

Figure 2.1 Basic concept illustration for Power-to-Gas (Adapted from Reference 2)

With this addition, the vast NG infrastructure can offer the electricity system a very large, distributed, energy storage network that can also transport energy from one area of the province, or nation, to another while shifting the time between generation and end-use from hours, days or months. In Figure 2.1, the system includes multiple sections: energy supply, energy conversion, the transmission and storage systems, distribution, conversion and final use. The hydrogen conversion technology, electrolysis, has the advantage of releasing no harmful emissions in comparison to steam methane reforming (SMR).

Two types of electrolyzers are available commercially for the production of hydrogen, alkaline and Polymer Electrolyte Membrane (PEM) electrolyzers. PEM electrolyzers are preferred for their ability to manage the intermittent and changing energy supplies that occur when renewable energy is used or when the electrolyzer is providing the variable load for ancillary services. The hydrogen produced using surplus off-peak power is fed directly to hydrogen customers or injected into the NG pipelines to create Hydrogen Enriched Natural Gas (HENG) which is then sold to NG customers, used to produce electricity with a Combined Cycle Gas Turbine, or fed to a separator to reconstitute a pure hydrogen stream for industrial or transportation use.

In locations where seasonal energy demand varies due to climate or where energy needs to be transmitted a long distance, such as in Canada, seasonal energy storage and efficient energy distribution are required. The production of electrolytic

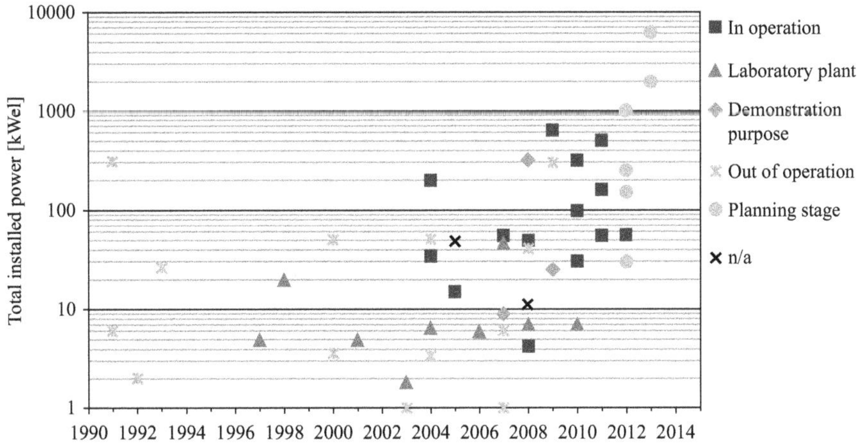

Figure 2.2 Increase Installed Power of Power-to-Gas Facilities [3]

hydrogen through PtG can provide the energy storage vector through the generation of hydrogen off-peak electricity, and then stored with NG in existing long-term underground storage facilities and NG pipelines. Note that the generation of the hydrogen can also be combined with bio-gas to capture CO_2. Due to the increasing prevalence of intermittent renewable energy sources, such as wind and solar power, and the increase in plug-in electric and fuel cell vehicles, the need for bulk energy storage, which can be distributed to meet demand, will only increase.

There are a number of PtG demonstration projects globally, with some producing hydrogen to be stored in tanks on-site and others using the NG infrastructure to distribute and store the gas [3]. In Figure 2.2, the increasing capacity of new PtG facilities with improving technology, is illustrated.

An example of a PtG plant using the NG pipelines is in Falkenhagen, Germany, which began operation in 2013 [4]. The plant, which was developed by E-ON in partnership with Hydrogenics, which provided the electrolyzer technology, is a 2 MW energy storage plant that uses surplus energy from renewable energy sources to produce hydrogen and injects it into a NG pipeline network. Under full operation the facility feeds approximately 360 Nm^3 per hour of hydrogen into the NG pipeline system. With the transition of the German power generation system to an increasing amount of renewable power, there are over 30 other demonstration projects in various stages of implementation [5]. The application of this system, in the context of Germany's energy system, is effective as it will help ensure a smoother transition toward their ambitious goal of meeting 50% of the energy demand by 2030 via renewable energy generation sources [6]. In Ontario, a 2 MW energy storage demonstration project for PtG is also under development [7].

2.1.1 Energy pathways in PtG

PtG offers numerous pathways which facilitate different energy storage, energy inputs and applications [8, 9]. In Figure 2.3, five unique pathways are given for PtG

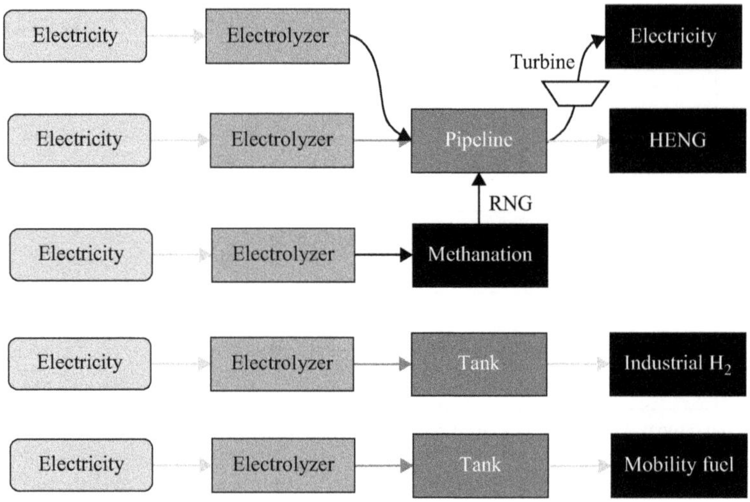

Figure 2.3 Different pathways for Power-to-Gas [10]

including: PtG-to-Electricity, PtG-to-Pipeline, PtG-to-Industrial Hydrogen, PtG-to-Methanation and PtG-to-Mobility.

The PtG-to-Pipeline path occurs when the hydrogen that is produced by electrolysis is injected into the NG infrastructure to create HENG. The HENG can be piped to NG end users, as long as the hydrogen does not take up more than 5% by volume. The PtG-to-Pipeline reduces emissions for NG and combined heat and power (CHP) customers [3]. PtG-to-Electricity, commonly called Power-to-Power, is a pathway whereby hydrogen is stored in a tank on-site to produce power on demand. Power-to-Power has a round-trip efficiency of only 40%, however, due to the extra energy conversion. PtG-to-Methanation refers to the production of hydrogen specifically for the methanation of CO and CO_2 to create Renewable Natural Gas (RNG). The RNG produced through methanation can then be injected into the NG pipelines. As discussed below, methanation is performed with the use of biogas from farm based digestion systems or landfill gas and is an excellent source of green energy. Due to the consumption of CO_2 in this pathway, PtG-to-Methanation can be seen as a type of carbon sequestration. These three pathways use the existing NG infrastructure to provide flexibility to the energy grid.

When hydrogen enters the NG infrastructure it can be considered blended with NG in the pipelines and thus used to generate electricity or be separated out for the last two applications, industrial H_2 and mobility fuel, close to the point of demand. PtG-to-Industrial Hydrogen is a preferred pathway of PtG for large industries, such as ammonium or petroleum production. Currently, the chemical industry consumes large amounts of hydrogen. If these industries seek to reduce the carbon intensity of their operations, due to the prevalence of carbon taxes or, as in Ontario, Cap and Trade, PtG-to-Industrial Hydrogen is an excellent technological choice. In this pathway, the green hydrogen can be delivered by pipeline or produced on-site, for industrial users who would otherwise consume hydrogen produced by SMR. The final PtG option, PtG-to-Mobility, occurs when the hydrogen produced through electrolysis is stored by tank for later use as a mobility fuel. Note that many fuel cell vehicles are expected to enter the market in the 2015–2020 timeframe [11].

2.2 Benchmarking PtG to other energy storage technologies

Due to the aforementioned increases in overall energy demand and the intermittency of newly developed renewable energy sources, an increasing number of energy storage technologies are being investigated for use with the current electrical infrastructure. A variety of applications can be improved through the utilization of energy storage including transportation, residential, commercial, industrial and utility-support projects. With the changing nature of energy demand, larger applications which are capable of supporting industrial and utility-scale storage needs are of specific interest. PtG can be utilized on a large enough scale to meet the aforementioned applications' needs; however, there are other advantages to PtG technology. The advantages to using PtG for these projects include the potential use of electrolytic hydrogen for fuel cell vehicles and the ability of electrolyzers to perform frequency regulation for utilities operators, increasing the flexibility of PtG applications and services delivered by the infrastructure. Approaches to the use of PtG for providing ancillary services for the electrical systems are discussed in this chapter.

A number of energy storage technologies already exist for industrial and utility-scale projects and have been analyzed by both American and European energy laboratories. These energy storage methods include pumped hydro, compressed air energy storage (CAES), flywheels, supercapacitors and a number of competing battery chemistries [12–14]. PtG, being a new energy storage technology, has not been assessed under these types of analysis.

In an examination of energy storage systems by Sandia National Laboratories, which utilize previous analyses by the Electric Power Research Institute (EPRI) [14], an analysis of the costs, maturity and performance of a selection of various energy storage technologies is performed [13]. The benchmarking method used by Sandia is useful for comparing the energy storage systems and can be extended to the investigation of energy storage using PtG. In the EPRI study, the various applications of PtG are grouped by the scale of capacity required. The proposed applications on the megawatt scale include bulk energy storage to support system and renewables integration; energy storage for ISO fast frequency regulation and renewables integration; energy storage for grid transmission and distribution (T&D) support applications; and energy storage for commercial and industrial applications [14].

2.2.1 *Comparing PtG with other energy storage technologies*

Comparative studies of energy storage technologies can be used to illustrate the strengths and weaknesses of the competing technologies. Ashby and Polyblank [15], for example, discuss multiple energy storage technologies and group them into hydrocarbons, mechanical storage, thermal storage, chemical storage and electromagnetic storage to meet a variety of applications. The International Energy Agency [16] also classifies energy storage projects by technology types including: pumped storage hydropower (gravitational energy), batteries (chemical energy) or flywheels (mechanical energy). Batteries themselves can be divided further based on their chemical composition and suitability for different applications [17].

To compare energy storage technologies with PtG, it is necessary to include criteria that characterize the different applications. The technical criteria from previous reports are thus combined with no criteria developed by Walker *et al.* [10]. The services considered are ancillary services for frequency with renewables integration, bulk energy storage, distributed energy storage, utility T&D support, commercial energy storage and residential energy storage [14, 18]. Each of these different applications requires unique properties of the energy storage system, as shown in Table 2.1.

Although a PtG system is able to provide each of the services listed in Table 2.1 separately, it is more likely that a system will be designed to provide multiple energy services, such as ancillary and bulk energy storage services. The bulk energy storage application, for example, can be accomplished by using the existing NG infrastructure to store hydrogen gas, while frequency regulation would be accomplished by varying the load of the electrolyzer to meet the demand requirements of the grid [20]. These types of systems, called energy hubs, are examined by Maroufmashat *et al.* [21] and Maniyali *et al.* [22], among others and are discussed later in this chapter. The performance of PtG for each of the applications presented in Table 2.1 is related to the ability of PtG to provide each function in comparison to other appropriate technologies. PtG is flexible enough to meet multiple applications simultaneously, however, and this is another advantage of the technology.

In the second column of Table 2.2, the 'Maturity' of PtG and other energy technologies using a qualitative evaluation which scores PtG between a speculative research idea and a fully-developed commercial technology [14]. As of the beginning of 2015, there are at least 35 PtG demonstration facilities operating in Europe which confirms this status [3, 19]. Given the number of facilities and maturity of PtG research, the technology best fits the 'demonstration' maturity category. The next group of criteria for comparing PtG with other energy storage technologies include: 'Maturity', 'Capacity', 'Power', 'Duration', '%Efficiency', 'Total Cost per kW' and 'Cost per kWh'. Adding to these criteria, shown in the shaded columns of Table 2.2, are additional criteria, including: 'Seasonal Storage', 'Portability', 'Energy Density', 'Potential for Mobility Fuel' and 'Consumer Acceptability'. The combination of these lists of criteria is used to evaluate the technologies for the six applications listed in Table 2.1.

Although some of the PtG service pathways perform the same against the different criteria, there are many differences between how frequency ancillary services and bulk storage would be handled by PtG. For bulk energy storage, the electrolyzer would operate either at 100% capacity, or whenever energy that is cheap or with a low-emissions factor is available. This hydrogen would then be stored in a tank, injected into the NG grid or injected into underground storage reservoirs. A key consideration which is advantageous for PtG technology is the

Table 2.1 Energy services applications for Power-to-Gas projects [10]

Service category	Requirements	Appropriate technologies	Power-to-Gas outlook
Ancillary services for IESO: voltage and frequency regulation, uninterrupted high power supply	Ancillary services typically encompass a large range of services and power ranges. However, for voltage and frequency regulation, a high energy density, high ramp speed and capacity are required.	Flywheels, superconducting magnetic energy storage, supercapacitors.	High ramp rates of electrolyzers used in Power-to-Gas can alter power operating level to offer voltage and frequency regulation services and this service has been demonstrated.
Bulk energy storage and renewables integration	Systems with high specific energy (kWh/kg) are given preference and long energy storage durations are preferred.	Pumped hydro storage, compressed air energy storage, flow batteries.	The ability of underground natural gas reservoirs to store large amounts of energy (close to 80 TWh) for long periods of time with minimal losses offers the potential to shift generation on a seasonal timescale. Bulk hydrogen generation can also be used for distribution to hydrogen refueling infrastructure [23].
Transmission and Distribution support for the electrical grid	Systems able to provide a high amount of power, energy and duration are preferred to meet peaks that are unable to be met by pre existing infrastructure.	CAES, flow batteries, advanced lead acid batteries, lithium ion batteries.	Power-to-Gas is able to provide energy to meet demands in remote location to its proficient transportation properties. In addition, the natural gas grid does not suffer from the distribution losses as significantly as the electrical grid.

(Continues)

Table 2.1 (Continued)

Service category	Requirements	Appropriate technologies	Power-to-Gas outlook
			Supports transmission grid congestion grid relief at peak periods.
Distributed energy storage systems	A system, likely taken on by utility with the ability to provide high efficiency storage and portability.	Advanced lead-acid batteries, flow batteries.	Power-to-Gas may provide solutions to matching demand and supply profiles by providing on-demand energy storage. Hydrogen production, storage and power generation do not have to be coupled in size or location. This can also support fleets or refueling stations.
Commercial and industrial energy storage	The system needs to be able to provide high voltage electricity, fuel for material handling or heating for a facility with a large square footage.	Advanced lead-acid batteries, flow batteries, lithium-ion batteries.	Hydrogen is a common industrial gas with many direct uses. Power-to-Gas may be used to generate hydrogen for fleets of hydrogen fuel cell vehicles, support mobility fleets, or to contribute to hydrogen enriched natural gas for heat or combined heat and power.
Residential energy storage	Able to provide either backup power or sufficient storage for daily peak shifting. High efficiency is important, in order for fiscal feasibility.	Advanced lead-acid batteries, flow batteries, lithium-ion batteries.	Power-to-Gas may be used to generate hydrogen for in-home refueling of hydrogen cars or to contribute to hydrogen enriched natural gas for heat or combined heat and power. Conversion back to electricity is also possible.

Table 2.2 Comparison of 'Power-to-Gas' with other energy storage technologies for large applications

Technology storage pathways	Maturity	Capacity (MWh)	Power (MW)	Duration range of storage (h)	Seasonal storage capability	%Efficiency	Total generation capacity cost ($/kW)	Energy storage cost ($/kWh)	Portability (% loss per 1000 km)	Energy density (kWh/m)	Mobility fuel	Consumer acceptability
Ancillary services (frequency) and energy storage for renewables integration												
Flywheel	Demo	5	20	0.25	No	91–96 [13, 14]	1950–2200	7800–8800	7	10–15	No	Accepted
Advanced lead-acid	Demo	0.25–0.5	1–100	0.25–1	No	75–90 [13, 14]	950–1590	2770–3800	7	30–90 [24]	No	Depends on scale
Li-ion	Demo	0.25–25	1–100	0.25–1	No	87–92 [25]	1085–1550	4340–6220	7	300–1000 [24]	No	Depends on scale
Bulk energy storage for renewables integration												
Pumped hydro	Mature	1680–5300	280–530	6–10	Yes (with losses)	80–82	2500–4300	420–430	7	0.7 [24]	No	Accepted
Compressed air	Comm.	1080–2700	135	8–20	No	71 (adiabatic), 25–40 [26]	1000–1250	60–125	7	2.4 [24]	No	Developing
Na/S	Comm.	300	50	6	No	75	3100–3300	520–550	7	100–200 [24]	No	Scale depend
Advanced lead-acid	Comm.	200	50	4	No	85–90	1700–1900	425–475	7	30–90 [24]	No	Scale depend
Vanadium	Demo	250	50	5	No	65–75 [27]	3100–3700	620–740	7	300–600 [24]	No	Scale depend
Zn/Br	Demo	250	50	5	No	60	1450–1750	520–550	7	300–600 [24]	No	Scale depend
Fe/Cr	R&D	250	50	5	No	75	1800–1900	520–550	7	300–600 [24]	No	Scale depend
Zn/air	R&D	250	50	5	No	75	1440–1700	520–550	7	500–800 [24]	No	Scale depend

(Continues)

Table 2.2 (Continued)

Technology storage pathways	Maturity	Capacity (MWh)	Power (MW)	Duration range of storage (h)	Seasonal storage capability	%Efficiency	Total generation capacity cost ($/kW)	Energy storage cost ($/kWh)	Portability (% loss per 1000 km)	Energy density (kWh/m)	Mobility fuel	Consumer acceptability
Power-to-Gas-Pipeline	Demo	7.84×10^5	1–100	7.84×10^3–10^5	Yes	85–90 [28]	~1000	~0.005	<0.001	4486 (H_2) [29]	Yes	Depends on scale
Power-to-Gas-Methanation	Demo	7.84×10^5	1–100	7.84×10^3–10^5	Yes	80 [28]	~1000	~0.005	<0.001	4486 (H_2) [29]	Yes NG cars	
Power-to-Gas-Electricity	Demo	7.84×10^5	1–100	7.84×10^3–10^5	Yes	~35	~1000 [26, 27]	~0.005	<0.001 [30]	4486 (H_2) [29]	No	Depends on scale
Power-to-Gas-Mobility	Demo	0.25–25	1–100	0.25–1	No	85–90 [28]	~1000	~0.005	<0.001	4486 (H_2) [29]	Yes	Growing
Power-to-Gas-Industrial Hydrogen	Demo	0.25–25	1–100	0.25–1	No	85–90 [28]	~1000	~0.005	<0.001	4486 (H_2) [29]	No	Common practice

Capacity: total amount of energy that can be stored, Power: total rated capacity, Duration Range of Storage: number of hours of energy available at the rated power, Seasonal Storage Capability: whether the technology has the capacity to store more than two months of energy, Efficiency: round-trip efficiency, Total Cycles: expected number of cycles over the product's life, Total Generation Capacity Cost: cost per kW of rated power, Energy Storage Cost: cost per kWh of potential energy storage, Portability: % energy loss per 1000 km of transport, Energy Density: amount of energy stored per unit volume, Potential Mobility Fuel: possibility of using the technology to propel vehicles, Consumer Acceptability: whether or not use of the technology is likely to desirable by consumers.

ability to use the NG distribution and storage infrastructure which has a significant capacity. In Ontario, for example, there are an estimated 155 billion cubic feet (bcf) of underground NG storage areas, the vast majority of them underutilized at this time [31, 32]. If hydrogen is estimated to take up 5% by volume, due to issues related to embrittlement of pipes the total volume available for storing hydrogen is 5% $\times$ 155 bcf = 7.75 bcf [33]. When using the energy density of hydrogen, this volume of hydrogen works out to approximately 78,400 MWh of energy.

2.3 Summary of current commercialization and research activities

This short-term future of PtG is bright due to the existing energy infrastructure that can be utilized to provide large-scale energy storage and ancillary services. The challenge for proponents of PtG is to develop the gas interchangeability standards and predictive system modeling that allow researchers, planners, system operators and industry to realize the full technical and economic potential. The response to this challenge is being led by industry which continues to develop new PtG facilities both in Europe and now in North America. As pointed out by Gahleitner [3], numerous PtG pilot plants exist, especially in Europe, utilizing the various energy flows discussed above. In Section 2.3.1, current research activities to investigate PtG at the University of Waterloo in Waterloo, Ontario, Canada are discussed.

To bolster the use of PtG by the electrical and NG energy suppliers, ongoing research into the use of PtG to integrate the two systems continues. The key areas of PtG research at the University of Waterloo are: Methanation, Urban Energy Hubs, System Optimization, Ancillary Services and Economic Modeling. These research activities utilize a team of faculty, post-docs, students and industrial partners to provide solutions and prescribe future operating practices.

2.3.1 Methanation/RNG

To meet ongoing energy needs, reduce greenhouse gas (GHG) emissions and provide flexibility to energy systems, the use of RNG and synthetic natural gas (SNG) can bolster the current, existing NG supply system while consuming captured wastes rich in GHGs. These greener NG streams, both of which can be utilized as part of a PtG architecture, are the focus of both the joint application for the development of a RNG market in Ontario and Alberta, Canada and the expansion of SNG, RNG and PtG in Germany [34, 35].

RNG, sometimes called bio-methane, is processed from an intermediate product called biogas or landfill gas. Both biogas and landfill gas result from the decomposition of organic material in an oxygen-free environment; however, biogas is produced in an anaerobic digester while landfill gas occurs due to the natural breakdown of organic waste in landfill facilities. The main products of anaerobic digestion are methane and CO_2. For the purposes of waste management, digesters are often used with facilities that produce organic waste, including farms, waste treatment plants, municipal waste sites where organics are collected separately, and breweries, food or beverage plants [36, 37]. As many of these waste streams naturally undergo anaerobic digestion which releases methane, a GHG much more damaging than carbon dioxide, it is useful to capture these emissions and process them for RNG. Often the definition of RNG can be expanded to include other types of methane that have their origin in renewable sources:

- methanation of CO_2 and electrolytic hydrogen generated by wind or solar power
- production by bacteria/fermentation, referred to as bio-methane
- methanation of biogas and electrolytic hydrogen generated by wind or solar power

If a biogas collector is used to power on-site large diesel engines or, in some cases gas turbines, the intermediate biogas may be used which produces more CO_2 and runs at a low efficiency. However, when the biogas is processed to create RNG, the product is a gas that is interchangeable with NG. This RNG can then be injected into the NG distribution or transmission system where it is transported to utility customers' homes and businesses where it is burned in existing heating, water heating and process equipment. RNG can produce full-cycle efficiencies of up to 80%, depending on the end-use NG equipment [38]. When biogas is collected, it can be added to H_2 produced electrolytically from renewable energy or excess base load to create RNG which can be inputted into NG pipelines. This process involves the creation of biogas, the use of electrolysis to produce hydrogen and methanation to create a product that can be safely transported with the NG.

SNG is formed from fuels or coal using a process including gasification and methanation. During the gasification step, the organics or fuel are exposed to a high pressure steam to form syngas, which is made up of CO, CH_4 and CO_2. Next, the syngas is cleaned of acid gases, particulate matter or other impurities. Finally, methanation is used to form SNG. In the methanation of SNG, the hydrogen is formed through SMR or a thermocatalytic process. Using these technologies for producing SNG and RNG, it is possible to meet ongoing energy needs while also reducing GHGs.

In Ontario, Canadian energy utilities developed a concept to use RNG within the NG infrastructure. The plan includes purchasing RNG from Ontario producers over a period of 20 years, developing a maximum annual volume cap, supporting the RNG infrastructure, including quality control and providing access to infrastructure for all producers, and a 5-year contract purchasing window. Further, their plan calls for the creation of a purchasing market, dependent on a total consumable cap of RNG production. Eventually, this could lead to a wholesale market for RNG similar to the current NG market. The introduction of green NG is a supply-side initiative that would reduce emissions without asking consumers to make any changes in their pattern of energy consumption. One of the most exciting elements of using RNG is that in Ontario, can cause a maximum near-term, potential reduction of GHG emissions. This reduction from RNG in Ontario is about 13 million tonnes of CO_2e per year, or more than 45% of Ontario's 2020 GHG emissions reduction target.

In the utilities' bid to start a RNG system, they noted the following four points as important for a sustainable system. First, there should be a manageable customer bill impact. A study of NG customers found, in fact, that the majority of customers are willing to support a 4% increase in the price of NG in order to bring RNG online. Secondly, there should be market transparency. Thirdly, there should be appropriate cost recovery so that producers and distributors are able to fiscally justify the capital expenditure. Finally, on a similar note there should be a return on investment for producers [38].

In Germany, the ambitious *Energiewende* has instilled a shift toward renewable energy sources. Part of this shift has meant embracing biogas and SNG as both carrier fuels and methods of storing energy through PtG and methanation. For example, Bajohr and Goetz [39] examine the application of methanation to PtG applications while Goetz *et al.* [40] and Lefebvre *et al.* [41] describe the use of bubble columns to improve the methanation process used to create SNG, while Mueller *et al.* [42] explore the reaction kinetics of the gasification step of creating SNG. As methanation, and thus the creation of SNG, can also be connected to the production of hydrogen through PtG the joint research venture seems an important next step.

Like Ontario, whose government has committed to eliminating coal powered plants and implementing Cap-and-Trade, Germany's *Energiewende* has had a large impact on national industry. As national utility companies were examining the introduction of a new national market for green NG, so is Germany. For example, German automakers like Audi and Mercedes are looking to create NG model vehicles due to the role of PtG in the new energy [43]. SNG production can be produced for the energy and transportation sectors to reduce solid waste and emissions while conserving resources and providing flexible power. This was the stimulus for the Werlte SNG plant and its accompanying novel concept [44].

2.3.2 Energy hubs and networks

The investigation of energy hubs, applications where a variety of energy generation and transfer technologies are implemented, illustrates the potential energy storage and energy efficiency that can be used to support the Smart Grid. When these hubs are combined as an energy network and allowed to exchange energy, the collective efficiency of the hubs increases, improving financial and environmental performance. The interconnectedness of the energy network design provides an added layer of reliability for each of the network members. Maroufmashat *et al.* [21] examine a complex network of energy hubs under different scenarios to determine the financial viability and reduction of GHG emissions of such a design.

2.3.2.1 Combined heat and power

The use of distributed power generation, as occurs when energy hubs are connected, is expanding with an increase in renewable energy and a move toward the creation of smart energy networks (SEN) [45]. These networks are a natural extension of the Smart Grid concept where all available energy resources from the electrical and NG grids are considered [46]. A SEN meets energy demands more effectively than traditional energy infrastructures by integrating renewable and non renewable energy generation technologies across different energy vectors, and by having the ability to access untapped reservoirs of energy. SENs may be used to provide the effective utilization of renewable energy which is necessary to address ongoing issues related to the growth in energy demand, climate change and the need for energy security [47]. The use of CHP systems as a component of a SEN, can improve its energy efficiency. In addition, the reliability of the network increases through the ability to generate electricity and heat during power outages to the entire network of energy hubs. Using this integrated approach, clean energy vectors such as hydrogen and RNG can also be used to store energy or service vehicle fleets.

2.3.2.2 Case studies of SENs

The two case studies that examine different SENs with two and three energy hubs within an overall energy network are considered. The modelling scenarios vary according to the consideration of distributed energy systems and energy interaction between energy hubs. In research carried out by Maroufmashat *et al.* [21], there is a significant increase in CO_2 emissions reductions and economic benefits when the number of hubs in the SEN increases. This demonstrates, perhaps, that there is a need for a diversity of loads and sources in a SEN to achieve significant benefits.

2.3.3 System optimization and ancillary

Balancing the difference between supply and demand of energy in an optimal way is a primary obligation of the power grid and thus optimizing the integrated energy system by providing ancillary services is an important application of PtG technology. Since the load placed on the electrical grid varies, the balancing authority procures and implements contingency reserves to maintain the reliability of the system. These services are split into two markets: the real-time energy market and the ancillary services market. The real-time market handles offers and bids from various suppliers and distributors of energy, while the ancillary market is made up of generators and responsive loads competing to offer back-up generation or decreased consumption when there is an imbalance between supply and demand.

Maintaining the reliability of the electrical grid is complex, especially due to the increased use of renewable energy. As the load met by the grid at night is close to the power produced by base load generators, which are mainly nuclear and hydroelectric power plants, and these plants cannot alter their power generation surplus base load generation occurs [48]. Without energy storage, the addition of renewable sources only exacerbates this issue. Currently, in locations like the province of Ontario, surplus energy is exported to neighboring jurisdictions at low and negative rates.

Although a number of PtG set-ups can be used to manage surplus base load and fluctuating renewable energy generation, interesting case studies performed at the University of Waterloo include: the use of wind energy to power an electrolyzer which produces hydrogen that is injected into the NG infrastructure and an optimization study where hydrogen is generated using electricity that would be exported to neighboring jurisdictions. Since Ontario has a summer peaking electrical grid, it is not uncommon to see periods of surplus power generation during the winters when intermittent renewables like wind farm units have a higher output. Therefore, the idea of converting this excess energy into hydrogen and storing it until there is a surge in demand during the summer is a possible solution to surplus generation. It is also important to develop energy storage for wind power generation in order to avoid incidents where wind power producers are forced to not deliver energy [49, 50].

When considering the storage of hydrogen in the NG system, Melaina *et al.* [51] suggest that it is safe to have hydrogen concentrations of up to 20% in NG pipelines, while other stress a limit as low as 5% [33]. To validate the feasibility of the PtG concept, the variation in the monthly average concentration of hydrogen when it is injected in to a constant high flow NG pipeline is also examined in both scenarios.

Both of the case studies necessitate solving a mixed integer linear programming (MILP) problem, where the outputs of the models are the optimal number of electrolyzers and storage units. A multi-objective approach is adopted to solve the two MILP problems. Each of the models has two objectives: (1) an Economic Objective and (2) an Environmental CO_2 Emission Objective. The outputs to the models include: optimal numbers of electrolyzers and storage tanks, dynamic hydrogen concentration within the distribution pipeline, costs of installing and operating the energy hub, emissions reductions, and power export savings.

2.3.4 Economic analysis of PtG systems

To examine the economic performance of PtG systems, a MATLAB® function can be used to examine different market mechanisms [52]. This is done in a two stage process outlined in Figure 2.4. First, a profit matrix is created in which the yearly base electricity cost, additional electricity fees, operational cost, and revenue from hydrogen production from 2011 to 2013. This matrix is then manipulated to determine the average yearly profit.

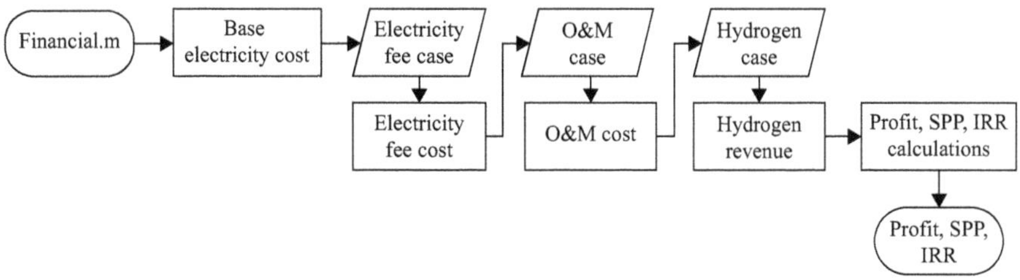

Figure 2.4 Logic for the calculation of financial viability of specific market mechanisms

To compare the different market mechanisms, in the context of Ontario, the base electricity price and maintenance costs are kept constant. Though the simulation is designed to accommodate any 5-year period where hourly data is available, only the hourly Ontario energy price (HOEP) is considered in this work while the operating and maintenance fees have been set to \$0.00446 per MWh [53, 54]. An additional electricity fee is charged to account for T&D charges. In 2012, a review of the tariff of \$2 per MWh for electricity exports resulted in a proposed tariff, called the EANC of \$5.80 per MWh [55]. As the electricity for the simulation is used internally to the jurisdiction the EANC is used as an additional market mechanism for the electricity fee cost.

Market mechanisms that treat hydrogen solely as an energy source and as a product are considered in this work. Utilizing Henry Hub data obtained from the US Energy Information Administration (EIA) for the monthly NG price for the examined time period hydrogen can be sold for its energy value. The monthly mmBTU is calculated for each year and the revenue from selling hydrogen for its energy value is calculated. Hydrogen can also be sold as a product on a mass basis. Utilizing data from the Department of Energy a current price of \$5.08 per kg is considered [56]. This price is used to see if a PtG system with this structure can be competitive with SMR hydrogen production and industrial hydrogen practices.

Hydrogen can also be sold for its renewable alternative energy value by examining the price of ethanol. The current price of ethanol is \$2.58 per gal which is equal to \$30.47 per mmBTU, in US dollars [56]. By selling the hydrogen at this price the energy value of both renewable alternative fuel sources is explored and compared. After the profit is determined, the capital costs and viability are determined. Capital costs are calculated based off of numbers given by the industrial partners. The capital cost is calculated by setting the low end cost to \$1.5 per MW and \$1.25 per MW installed for a 5 and a 40 MW plant respectively, in US dollars. The capital cost is then interpolated to obtain the investment for the in between plant sizes. Utilizing this profit matrix and the capital cost the function calculates the simple payback period (SPP) and the internal rate of return (IRR). A lifetime of 10 years is used in this work to calculate the IRR [54]. The results are then used to compare the viability of the different mechanisms summarized in Table 2.3.

Table 2.3 Market mechanisms examined in financial simulations

Parameter	Market mechanism	Value (all \$ Canadian)
Electricity base price	Hourly Ontario electricity price	IESO hourly data [53]
Electricity fees	Equivalent average network charge	\$5.80 per MWh [55]
Operation and maintenance	Operation and maintenance excluding electricity	\$4.5 $\times$ 10^{-3} per MWh [54]
Revenue	Natural gas energy price	EIA Henry Hub Monthly data [57]
	Industrial hydrogen price (SMR)	\$4.68 per kg [56]
	Ethanol energy price	\$330.90 per mmBTU [56]

2.3.4.1 Results of the economic analysis

The market mechanism viability is determined from the SPP and the IRR. Since the NG energy price is much lower than the SMR and ethanol prices, it can be eliminated from the analysis. As both the SMR and ethanol market values produce positive profits they are examined for their viability.

The SPPs for operating the electrolyzers at a low operating capacity are higher than the expected lifetime of the electrolyzers themselves. This makes a low operating capacity factor not a viable option as the plant would not make its return on investment within 10 years. So although the market mechanism seems viable, it cannot be used. Operating a high operating capacity, above 90%, however, creates a substantial profit. In this case, the SPP ranges from 5 years at low power capacities to 4 at high power capacities.

2.3.4.2 Further discussion

In addition, PtG facilities show potential to provide substantial benefits to improve the operation of power grids by providing increased grid management as well as important auxiliary services. The market mechanism for how a facility that generates hydrogen in a PtG system operates has not yet been determined, but it is important to examine to demonstrate the economic viability of PtG systems. Though hydrogen cannot be sold efficiently for its energy value at the NG price, other mechanisms that compare the hydrogen product with SMR hydrogen and ethanol show that the projects can be viable when compared these energy vectors. This is only true however if higher capacity factors are used, meaning the facilities focus cannot solely be used for the reduction of energy exports. The electrolyzer technology must be used for energy arbitrage and provision of ancillary electrical services. Even great economic benefits can be realized with the sale of hydrogen into industrial or mobility markets.

Capital incentives can also be used to increase the profitability of PtG systems. These incentives tackle directly the obstacle of large capital cost of these systems. Currently in the United States, there are multiple programs that can be used to cover up to 80% of the capital cost of alternative fuels as long as a portion of the funding goes toward refueling infrastructure. This program would be extremely beneficial to the distributed refueling infrastructure discussed earlier. Thought a capital incentive is a large sum, it forces the company to continuously improve its production and operation as to maintain profitability. It removes the risk of government withdrawal of the program and an end to funding as with production incentives.

2.3.5 Cap-and-Trade

In this research, the use of electrolytic hydrogen to generate earnings through Cap-and-Trade emissions programs is investigated [58]. Due to the high demand for industrial hydrogen, substituting electrolytic hydrogen for hydrogen from the more commonly used SMR for industrial uses, such as petroleum refining, represents the best opportunity to reduce GHG emissions considering current hydrogen use pathways. Simons and Bauer [59] find that if a grid mixture dominated by nuclear energy is used to power electrolyzers, approximately 3 kg of CO_2e is emitted per kg of H_2 are produced by electrolysis, while the SMR processes emit about 15 $kgCO_2e/kgH_2$.

The hydrogen produced by PtG can be considered as a 'green' fuel as it is making use of excess nuclear and intermittent renewable power. As the technology providing critical energy storage to enable the Smart Grid, hydrogen produced via PtG should be compared on an energy pricing basis to other 'renewable' and 'green' fuels such as ethanol and bio-diesel. To encourage reductions in CO_2 emissions, a number of carbon reduction systems have been proposed. Carbon taxes have been proposed whereby organizations are made to a pay a premium for the carbon emissions that they are responsible for [60, 61]. One criticism of the carbon tax, however, is that it generally increases the costs of energy and transportation for consumers without encouraging innovation [14, 60]. Cap-and-Trade systems, however, have been successfully implemented in the European Union and North America [62, 63]. The Cap-and-Trade systems work on the premise that companies that are able to reduce their CO_2 emissions to lower levels should be able to reap financial benefits. It does this by setting a 'hard cap' of global emissions from a region or nation and for individual businesses [64]. Under the hard cap, companies that reduce their emissions below the hard cap are given permits that they are able to sell to other companies. Companies that are unable to meet the hard cap must purchase the permits from the market. The importance of carbon trading systems cannot be understated. By monetizing GHG emissions, technologies that may be more expensive, but are capable of providing carbon reductions, Cap-and-Trade encourages technological improvements. As the province of Ontario transitions into a Cap-and-Trade system for industry, the analysis of how such a system could be used for hydrogen production by PtG is of current interest [65].

PtG facilities show potential to provide substantial benefits to improve the operation of power grids by providing increased grid management as well as important auxiliary services. The market mechanisms for how a facility that generates hydrogen in a PtG system have not yet been determined, but are an important to examine to demonstrate the economic viability of PtG systems. Though hydrogen cannot be sold efficiently for its energy value at the NG price, other mechanisms that compare the hydrogen product with industrial SMR hydrogen as well as renewable ethanol price show the projects can be viable when compared these energy vectors. This is only true however if higher capacity factors are used, meaning the facilities focus cannot solely be used for the reduction of energy exports. The electrolyzer technology must be used for energy arbitrage and provision of ancillary electrical services in order to be competitive. Even greater economic benefits can be realized with the sale of hydrogen into industrial or mobility markets. With high capacity factors and appropriate market mechanisms IRR between 15% and 21% are achieved demonstrating the project viability when the energy value of hydrogen is compared to that of ethanol energy. In addition to offering a high capacity for energy storage, PtG also provides industrial hydrogen producers the opportunity to increase their profits through the sale of carbon credits. However, it is also seen that carbon credits for hydrogen producers can amount between 7% and 10% of the total costs to purchase and operate the plant. Because additional electricity tariffs are so significant, and the SPPs are reasonable for lower capacities, it is most reasonable to develop a 5 MW system operating with a switch price of $30 per MWh. Under such plan, the lower fees would apply and the plan is profitable. As the commodity pricing of hydrogen is more useful than the energy price of hydrogen, it is reasonable to assume that the producer would prefer to sell the hydrogen for petroleum refining, as opposed to selling it for transportation fuel at this time. With high capacity factors, set prices and appropriate market mechanisms, IRR between 20% and 30% are shown, which demonstrates the viability of using Cap-and-Trade to compare hydrogen production via PtG with that of SMR pricing for electrolytic hydrogen. An advantage to supporting this for policy maker organizations and government is that PtG encourages the convergence of the NG and electrical infrastructure, achieves energy management benefits, and improves the production of hydrogen to support a future hydrogen economy. These mechanisms should be implemented by governments and policy maker organizations in order that convergence of the NG and electrical infrastructure can achieve energy management benefits, as well as other benefits including energy storage, clean industrial hydrogen and future development of a clean hydrogen mobility infrastructure.

2.4 Summary

There continues to be growing interest in the development of PtG technology to provide energy storage, ancillary services and to provide cleanly produced hydrogen fuel for future generations of vehicles. To grow the industry, continuing research will be undertaken into using Cap-and-Trade or carbon taxes to provide financial incentives; utilizing green hydrogen for emissions savings in the production of petroleum and other products; the utilization of electrolysis systems to provide ancillary services to grid operators; and the use of the NG infrastructure to provide the long-term bulk storage needed to manage excess base load and integrate renewable power sources.

References

[1] M. Fowler and U. Mukherjee, "Adaptive energy ecosystems," *IEEE Canadian Review*, vol. Spring, pp. 20–25, 2014.

[2] U. Mukherjee, M. W. Fowler and A. Hajimiragha, "Power-to-Gas energy storage," *International Journal of Hydrogen Energy*, vol. 40, no. 31, pp. 976–991, 2015.

[3] G. Gahleitner, "Hydrogen from renewable electricity: an international review of Power-to-Gas pilot plants for stationary applications," *International Journal of Hydrogen Energy*, vol. 38, no. 5, pp. 2039–2061, 2013.

[4] Hydrogenics, "*Falkenhagen Power-to-Gas plant,*" 2013. [Online]. Available: http://www.hydrogenics.com/about-the-company/news-updates/2013/06/14/largest-Power-to-Gas-facility-in-the-world-now-operational-with-hydrogenics-technology. [Accessed 22 January 2015].
[5] Germany Trade and Invest, "*Power-to-Gas technology,*" 2015. [Online]. Available: http://www.gtai.de/GTAI/Navigation/EN/Invest/Industries/Smarter-business/Smart-energy/power-to-gas,t=powertogas-strategy-platform,did=589148.html. [Accessed 12 September 2015].
[6] (dena), Deutsche Energie-Agentur GmbH, "*Deutscher Energiewende-Index,*" 4 October 2013. [Online]. Available: http://www.dena.de/projekte/energiesysteme/deutscher-energiewende-index.html. [Accessed 1 May 2015].
[7] Hydrogenics, "*Hydrogenics selected for 2 megawatt energy storage facility in Ontario,*" 25 July 2014. [Online]. Available: http://www.hydrogenics.com/about-the-company/news-updates/2014/07/25/hydrogenics-selected-for-2-megawatt-energy-storage-facility-in-ontario. [Accessed 22 September 2015].
[8] M. Specht, M. Sterner, F. Baumgart, et al., "*New routes for the production of substitute natural gas (SNG) from renewable energy,*" ForschungsVorbund Ernueurbare Energien, Germany, 2010.
[9] M. Sterner, "*Bioenergy and renewable power methane in integrated 100% renewable energy systems,*" Kassel University Press, Kassel, 2009.
[10] S. Walker, U. Mukherjee and M. Fowler, "Benchmarking and selection of Power-to-Gas for energy storage," in *International Conference for Hydrogen Production (ICH2P)*, Oshawa, 2015.
[11] Pacific Northwest National Laboratory, "*Pathways to commercial success: technologies and products supported by the Fuel Cell Technologies Office,*" U.S. Department of Energy, Richland, 2015.
[12] P. Mannan, G. Baden, L. Olein, et al., "*Comparison of batteries, compressed air and Power-to-Gas energy storage technologies in the Alberta context,*" Alberta Innovates – Technology Futures, Edmonton, 2014.
[13] Sandia National Research Laboratories, "*Electricity storage handbook in collaboration with NCREA,*" DOE/EPRI, Sandia, NM, 2013.
[14] Electric Power Research Institute, "*Electricity energy storage technology options,*" Palo Alto, CA, 2010.
[15] M. Ashby and J. Polyblank, "*Materials for energy storage systems – a white paper,*" Granta Design, University of Cambridge, Cambridge, UK, 2012.
[16] International Energy Agency, "*Energy storage technology roadmap,*" 2014. Retrieved May 7, 2015, from: https://www.iea.org/publications/free publications/publication/TechnologyRoadmapEnergystorage.pdf.
[17] K. Divya and J. Ostergaard, "Battery energy storage technology for power systems—an overview," *Electric Power Systems Research*, vol. 79, no. 4, pp. 511–520, 2009.
[18] McKinsey & Company, "*Commercialization of energy storage in Europe,*" Fuel Cell and Hydrogen Joint Undertaking, Brussels, Belgium, 2015.

[19] Hydrogenics, Interviewee, "*Phone conversation.*" [Interview]. 15 February 2015.
[20] D. Wilson and D. Teichroeb, "*Power-to-Gas: utility-scale energy storage,*" IESO, 2012. Retrieved January 20, 2015, from: http://www.ieso.ca/Documents/smart_grid/materials/20120605/SGF-20120605-Hydrogenics.pdf.
[21] A. Maroufmashat, A. Elkamel, M. Fowler, S. Sattaria, A. Hajimiragha and E. Entcheve, "Modeling and optimization of a network of energy hubs to improve economic and emission considerations," *Energy*, vol. 93, No. 2, pp. 2546–2558, 2015.
[22] A. Maniyali, A. Almansoori, M. Fowler and A. Elkamel, "Energy hub based on nuclear energy and hydrogen energy storage," *Industrial & Engineering Chemistry Research*, vol. 52, no. 22, pp. 7470–7481, 2013.
[23] A. Hajimiragha, C. Canizares, M. Fowler, M. Geidl and G. Andersson, "Optimal energy flow of integrated energy systems, with hydrogen economy considerations," in *Bulk Power System Dynamics and Control – VII, August 19–24, 2007*, Charleston, 2007.
[24] T. Bauer, *"Thermophotovoltaics: basic principles and critical aspects of system design,"* Springer Science, New York, NY, 2011.
[25] L. Valoen and M. Shoesmith, "The effect of PHEV and HEV duty cycles on battery and battery pack performance," in *PHEV 2007*, Winnipeg, 2007.
[26] P. Denholm and G. Kulcinski, "Lifecycle energy requirements and greenhouse gas emissions from large scale energy storage systems," *Energy Conversion and Management*, vol. 45, pp. 2153–2172, 2004.
[27] Electric Power Research Institute, "*Vanadium redox flow batteries: an in-depth analysis,*" Palo Alto, CA, 2007.
[28] National Renewable Energy Laboratory, "*Technology brief: analysis of current-day commercial electrolyzers,*" U.S. Department of Energy, 2004.
[29] D. Green and R. Perry, "*Perry's chemical engineering handbook.*" 8th Edition, McGraw-Hill Education, New York City, NY, 2015.
[30] K. Mazloomi and C. Gomes, "Hydrogen as an energy carrier: prospects and challenges," *Renewable and Sustainable,* vol. 16, no. 5, pp. 3024–3033, 2012.
[31] Ontario Ministry of Natural Resources, "*Underground storage-reservoirs,*" Queen's Press, Toronto, 2012.
[32] D. Peng, *Enabling utility-scale electrical energy storage through underground hydrogen-natural gas co-storage. Master's Thesis*, University of Waterloo, Waterloo, 2013.
[33] K. Altfield and D. Pinchbeck, "Admissible hydrogen concentrations in natural gas systems," *Gas for Energy (German)*, vol. 13, pp. 1–12, 2013.
[34] Canadian Gas Association, "*Roadmap for renewable natural gas technology roadmap for Canada,*" Government of Canada, Ottawa, 2014.
[35] D. E.-A. G. (dena), "*Roadmap for Power-to-Gas,*" 2015. [Online]. Available: http://www.powertogas.info/plattform/information-in-english/roadmap-for-Power-to-Gas.html Accessed 26 May 2015. [Accessed 12 June 2015].

[36] M. Chutumstid, J. Woo, B. Yu, M. Fowler and M. Pritzker, "Integrated energy systems for agricultural applications," in *Fourth Year Poster Session*, Waterloo, 2013.
[37] (dena), Deutsche Energie-Agentur GmbH, "*Biogas production*," 2015. [Online]. Available: http://www.renewables-made-in-germany.com/en/renewables-made-in-germany/technologies/biogas.html. [Accessed 15 April 2015].
[38] Utilities, "*Renewable natural gas application*," 2011.
[39] S. Bajohr and M. Goetz, "Development of a methanation process for PtG applications," in *EDGar/DVGWM Conference, February 7–8, 2013*, Arnhem, 2013.
[40] M. Götz, S. Bajohr, F. Graf, R. Reimert and T. Kolb, "Einsatz eines Blasensäulenreaktors zur Methansynthese," *Chemie Ingenieur Technik*, vol. 85, no. 7, pp. 1146–1151, 2013.
[41] J. Lefebvre, M. Götz, S. Bajohr, R. Reimert and T. Kolb, "Improvement of three-phase methanation reactor performance for steady-state and transient operation," *Fuel Processing Technology*, vol. 132, pp. 83–90, 2015.
[42] A. Mueller, P. Stoesser and T. Kolb, "Biomass char gasification: study on reaction kinetics using a high pressure thermogravimetric analyzer," in *Clearwater Energy Conference*, Edmonton, Alberta, 2014.
[43] Q. Schiermeier, "Renewable power: Germany's energy gamble. An ambitious plan to slash greenhouse-gas emissions must clear some high technical and economic hurdles," *Nature*, 2013. [Online]. Available: http://www.nature.com/news/renewable-power-germany-s-energy-gamble. [Accessed 7 June 2015].
[44] NTVA, "*Etogas delivesr world's largest methane production plant to Audi*," 25 June 2014. [Online]. Available: http://www.ngvaeurope.eu/etogas-delivers-worlds-largest-methane-production-plant-to-audi. [Accessed 3 June 2015].
[45] M. Akorede, H. Hizam and E. Pouresmaeil, "Distributed energy resources and benefits to the environment," *Renewable Sustainable Energy Review*, vol. 14, no. 2, pp. 724–734, 2010.
[46] I. Rowland, "*Smart energy network: concept development*," Waterloo Institute of Sustainable Energy, November 2011. [Online]. Available: http://www.questcanada.org/publications/icesliteracyseries1. [Accessed 1 August 2014].
[47] Quest, "*Building smart energy communities: implementing integrated community energy solutions*," September 2012. [Online]. Available: http://www.questcanada.org/sites/default/files/publications/Building%20Smart%20Energy%20Communities%20-%20Implementing%20ICES.pdf. [Accessed 1 August 2014].
[48] Ontario Society of Professional Engineers, "*Wind and the electrical grid – mitigating the rise in electricity rates and greenhouse gas emissions*," Toronto, 2012.

[49] Energy Information Administration, "*Fewer wind curtailments and negative power prices seen in Texas after major grid expansion*," 24 June 2014. [Online]. Available: http://www.eia.gov/todayinenergy/detail.cfm?id=16831#. [Accessed 16 October 2015].

[50] E. Smith, "*Ratepayers paying wind farmers not to produce electricity — $2.7 million so far this year – red flag for region as I-937 kicks in – oversupply payments, half from BPA customers, expected to grow as new wind farms go online*," 12 September 2012. [Online]. Available: http://washingtonstatewire.com/blog/too-much-windpower-rivers-surged-this-summer-andoversupply-cost-2-7-million/. [Accessed 16 October 2015].

[51] M. Melaina, O. Antonia and M. Penev, "*Blending hydrogen into natural gas pipeline networks: A review of key issues*," National Renewable Energy Laboratory, Golden, CO, 2013.

[52] D. van Lanen, S. B. Walker and M. Fowler, "Economic analysis of market mechanisms for Power-to-Gas," in *International Conference on Hydrogen Production*, Oshawa, 2015.

[53] IESO, "*Index of public reports*," 2015. [Online]. Available: http://reports.ieso.ca/public/. [Accessed 10 February 2015].

[54] B. Zakeri and S. Syri, "Electrical energy storage systems: a comparative life cycle cost analysis," *Renewable and Sustainable Energy Reviews*, vol. 42, pp. 569–596, 2015.

[55] Charles River Associates, "*Export Transmission Service (ETS) tariff study*," IESO, Toronto, 2012.

[56] U.S. Energy Information Administration, "*Annual energy outlook 2015*," 2015. [Online]. Available: http://www.eia.gov/beta/aeo/#/?id=12-AEO2015. [Accessed 1 May 2015].

[57] U.S. Energy Information Administration, "*Henry hub natural gas spot price*," 2014. [Online]. Available: http://www.eia.gov/dnav/ng/hist/rngwhhdm.htm. [Accessed 1 May 2015].

[58] S. B. Walker, D. van Lanen, M. Fowler and U. Mukherjee, "Economic analysis with respect to Power-to-Gas energy storage with consideration of various market mechanisms," in *International Conference on Hydrogen Production*, Oshawa, 2015.

[59] J. Simons and T. Bauer, "*Lifecycle assessment of hydrogen production*," Laboratory for Energy Systems Analysis, Paul Scherrer Institute, Villigen-PSI, 2010.

[60] J. Poterba, "*Tax policy to combat global warming: on designing a carbon tax*," National Bureau of Economic Research, Cambridge, MA, 1992.

[61] G. Metcalf, "Designing a carbon tax to reduce U.S. greenhouse gas emissions," *Review of Environmental Economics and Policy*, vol. 3, no. 1, pp. 63–83, 2008.

[62] European Commission, "*The EU Emissions Trading System (EU ETS)*," 2015. [Online]. Available: http://ec.europa.eu//clima/policies/ets/documentation_en.htm. [Accessed 14 May 2015].

[63] California Environmental Protection Agency, "*Cap and trade program*," 2015. [Online]. Available: http://www.arb.ca.gov/cc/capandtrade/capandtrade.htm. [Accessed 14 May 2015].
[64] Environmental Protection Agency, "*Cap and Trade*," 2009. [Online]. Available: http://www.epa.gov/captrade/basic-info.html. [Accessed 14 May 2015].
[65] Ontario Government, "*Cap and trade system to limit greenhouse gas pollution in Ontario*," 2015. [Online]. Available: http://news.ontario.ca/opo/en/2015/04/cap-and-trade-system-to-limit-greenhouse-gas-pollution-in-ontario.html. [Accessed 15 May 2015].

Chapter 3

Methanation for Power-to-Gas applications

Thomas Kolb[1], Siegfried Bajohr[1], Jonathan Lefebvre[1], Manuel Götz[2] and Dominik Schollenberger[1]

3.1 Methanation for Power-to-Gas applications

Methane can be produced via CO (3.1) and CO_2 methanation (3.2). However, the CO_2 methanation is more relevant in the context of Power-to-Gas (PtG) applications. The two methanation reactions are accompanied by further reactions such as the reverse water gas shift reaction (3.3) and the Boudouard reaction (3.4). Looking at the overall stoichiometry the CO_2 methanation can be seen as the combination of the CO methanation with the reverse water-gas shift. The Boudouard reaction producing unwanted carbon deposits on methanation catalysts is a big challenge especially for the CO methanation but of minor importance for the CO_2 methanation [1–5] and therefore PtG applications.

$$CO_2(g) + 4H_2(g) \rightleftharpoons CH_4(g) + 2H_2O(g) \rightarrow \Delta H_r^0 = -165.1\ \mathrm{kJ/mol} \tag{3.1}$$

$$CO(g) + 3H_2(g) \rightleftharpoons CH_4(g) + H_2O(g) \rightarrow \Delta H_r^0 = -206.3\ \mathrm{kJ/mol} \tag{3.2}$$

$$CO_2(g) + H_2(g) \rightleftharpoons CO(g) + H_2O(g) \rightarrow \Delta H_r^0 = +41.2\ \mathrm{kJ/mol} \tag{3.3}$$

$$2CO(g) \rightleftharpoons C(S) + CO_2(g) \rightarrow \Delta H_r^0 = -172.5\ \mathrm{kJ/mol} \tag{3.4}$$

Both CO and CO_2 methanation are highly exothermic reactions. Hence, high temperatures limit the CO and particularly the CO_2 conversion. As illustrated in Figure 3.1, a temperature below 225 °C (at 1 bar) or 300 °C (at 20 bar) is required to reach a CO_2 conversion of at least 98%. Figure 3.1 also illustrates the positive effect of pressure on the CO_2 conversion. Consequently, methanation reactors should be operated at high pressures in order to reach high methane yields.

[1]Karlsruhe Institute of Technology, Engler-Bunte-Institute, Fuel Technology, Engler-Bunte-Ring 1, 76131 Karlsruhe, Germany
[2]DVGW Research Center at the Engler-Bunte-Institute of the Karlsruhe Institute of Technology (KIT), Engler-Bunte-Ring 1, 76131 Karlsruhe, Germany

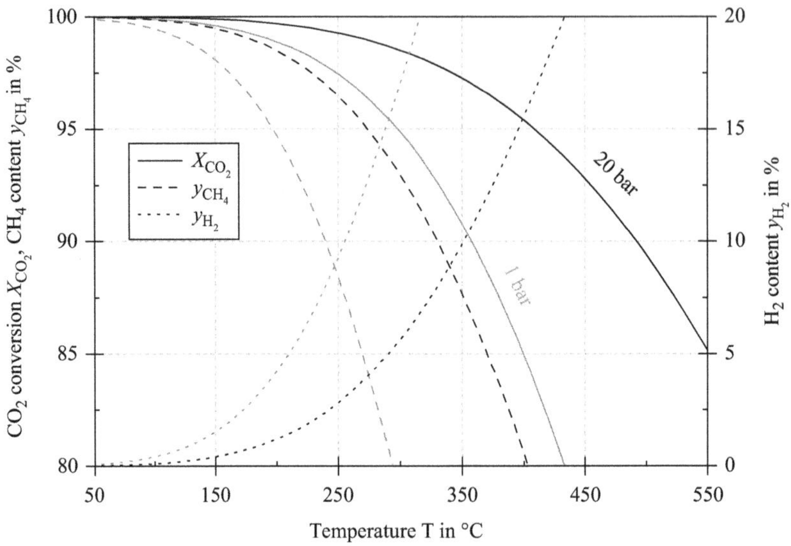

Figure 3.1 Equilibrium conversion of CO_2 as well as H_2 and CH_4 content for CO_2 methanation ($(H_2/CO_2)_{in} = 4$, no inert gases)

CO_2 methanation on commercially available methanation catalysts shows typically a selectivity of almost 100% [6–12]. Nevertheless, the calorific value of SNG produced from CO_2 methanation may be lower than the calorific value of natural gas as higher hydrocarbons typically found in the latter are lacking.

Figure 3.2 shows that a CO_2 conversion of almost 98% is required to achieve a methane content of >90%, whereas a CO_2 conversion of 99% correlates with a methane content of 95%. Inert gases or hyperstoichiometric H_2/CO_2 ratios make it impossible to achieve high methane contents. As a consequence, the requirements for the methanation reactor are very high.

3.2 Current research activities

One of the major drawbacks of PtG is its relatively low efficiency of methane production. Assuming a commercially available electrolysis technology delivering H_2 at 25 bar with an electrical efficiency of 70% and assuming a CO_2 methanation reactor operated at 20 bar with an efficiency of 78% (maximum chemical efficiency), an overall efficiency of 55% can be reached (Figure 3.3). A further efficiency loss of 2% results if CO_2 needs to be compressed to 20 bar. As a consequence, research focuses on an increase in PtG efficiency.

The Sankey diagram shows two improvement potentials for PtG. First, the efficiency of water electrolysis could be improved. Second, heat from the methanation reactor could be used.

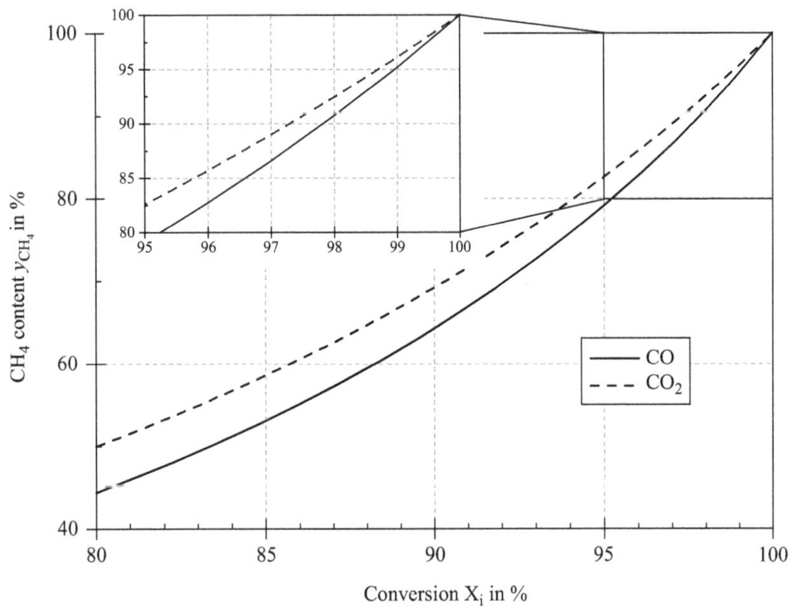

Figure 3.2 Relationship between CO or CO_2 conversion and methane content in the dry SNG (no inert gas; $H_2/CO = 3$ and $H_2/CO_2 = 4$, respectively; 100% selectivity)

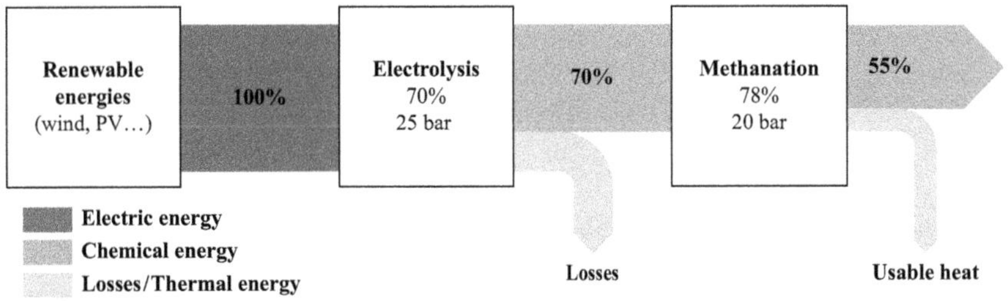

Figure 3.3 Sankey diagram of the PtG process efficiency (heat integration is not taken into account)

A stark amount of high temperature heat is produced during catalytic methanation and can be used, e.g. to produce steam in order to improve the PtG process efficiency. A methanation reactor producing 500 m^3/h of methane also generates more than 1 MW of heat. This amount of heat is at the same time the most significant issue for catalytic methanation reactors. Therefore, current research on methanation reactors focuses on how to realise an efficient reactor temperature control and heat extraction under the dynamic operating conditions of the PtG process.

3.3 Development of electrolysis and methanation technologies for PtG

3.3.1 Development of electrolysis technologies

Three different electrolysis technologies are under investigation for PtG process chains: alkaline electrolysis (AEL), polymer electrolyte membranes (PEM) and solid oxide electrolysis (SOEC).

3.3.1.1 Alkaline electrolysis

AEL is the most mature and well understood electrolysis technology which is commercially available for decades [13].

According to manufacturers, AEL electrolysers can be operated between 20 and 100% of their design capacity, and overload operation up to 150% is possible for short times. However, Gahleitner [14] reports problems with intermittent and fluctuating power sources. One problem is that it takes 30–60 min to restart the system following a shutdown [15]. Continuous operation of AEL systems is advised [15]. The biggest disadvantage of AEL is that the utilised electrolytes (alkaline solution, e.g. 20–30% potassium hydroxide solution) are highly corrosive, thus necessitating high maintenance costs. A general overhaul of the system is necessary every 7–12 years [16].

3.3.1.2 Polymer electrolyte membrane electrolysis

Compared to AEL, PEM is a relatively new technology, with the first commercial PEM electrolyser available for purchase in 1978 [17]. According to Reference 14, the main advantages of PEM include faster cold start, higher flexibility and better coupling with dynamic and intermittent systems. Furthermore, the purity of the produced H_2 is very high [18]. The minimum load is reported to be 5% [15, 16]. However, this technology is currently more expensive than AEL systems (due to the costs for the membrane and the use of a noble metal catalyst), and the limited life expectancy is also disadvantageous [18–22]. PEM electrolysers in the MW scale have recently become available (Proton Onsite and Siemens) [23].

3.3.1.3 Solid oxide electrolysis

SOEC (also known as high temperature electrolysis) is the most recently developed electrolysis technology and is still at the laboratory stage. The high temperature reduces the equilibrium cell voltage and therefore the electricity demand [24]; however, the heat demand increases with increasing temperature. The low electricity demand is the most significant advantage of SOEC systems. In the endothermic mode, electrical efficiencies above 100% could be achieved as part of the electrical energy demand is substituted by thermal energy [24, 25]. In particular, the combination with exothermal reactions in PtG and PtL (Power-to-Liquid) process chains leads to a higher overall efficiency [22, 26–30].

The biggest challenges for SOEC systems are the fast material degradation and the limited long-term stability, both of which are due to high temperature operation [17].

Moreover, SOEC systems are not stable against fluctuating and intermittent power sources [16, 17].

3.3.2 Development of catalytic methanation technologies

A significant issue in a methanation reactor is the realisation of a good temperature control in the reactor. In order to meet this essential objective, several steady-state reactor concepts have been developed, namely fixed-bed, fluidised-bed, three-phase and structured reactors. Fluidised-bed reactors as well as fixed-bed reactors are established technologies, while the other reactor concepts are in the development phase. The development of these methanation reactor concepts has already been reviewed elsewhere [31–33]. Each of these reactor concepts offers a different approach to overcome the issue related to the reaction heat removal (Table 3.1).

For **adiabatic fixed-bed reactors**, the usual approach relies on a series of adiabatic reactors, typically 2–5, with intercooling and sometimes with gas recirculation [35, 36]. Due to the adiabatic mode of operation, the catalyst must be able to withstand a broad temperature range (250–700 °C). The main concerns for the catalyst are possible cracking or sintering [2]. Alternatively, **cooled fixed-bed reactors** can be applied for methanation. Usually, such a reactor contains cooling tube bundles; a further possibility is the use of cooled plates [22, 31, 37]. Due to the cooling, the process setup is simpler than for adiabatic reactors, however, the reactor itself is more complex and consequently expensive.

In **fluidised-bed reactors**, the mixing of fluidised solids leads to almost isothermal conditions in the reactor facilitating the control of the operation. Offering more effective heat removal is the major advantage of this concept, which allows for using one single reactor with a rather simplified design [32, 38, 39]. Nevertheless, due

Table 3.1 Comparison of different methanation technologies

Reactor concept	**Adiabatic fixed-bed**	**Fluidised-bed reactors**	**Three-phase reactors**	**Structured reactors**
Operation mode	Adiabatic	≈ Isothermal	Isothermal	Polytropic
Technology readiness level [34]	9	7	4–5	5–6
Temperature range in °C	250–700	300–400	300–350	250–500
Stages	2–5	1–2	1–2	1–2
Gas recycle	Sometimes	No	No	No
Reactor costs	Low	Low–medium	Low–medium	High
Complexity of process setup	High	Low–medium	Low–medium	Low
Main advantage	High reaction rate	Simple process setup	Simple process setup	Temperature control
Main drawback	Complex process setup	Mechanical load on catalyst	Mass-transfer limitation	Reactor costs

to high mechanical load resulting from fluidisation, attrition processes take place in relation to the catalyst as well as the wall of the reactor. As a consequence, the catalyst is eventually deactivated [2]. A further disadvantage can be the incomplete CO_2 conversion caused by bubbling. In addition, a fluidised-bed reactor is limited by superficial gas velocity within the reactor: it cannot be too low in order to assure minimum fluidisation conditions and cannot be too high in order to avoid catalyst elutriation.

Other concepts are based on the use of **three-phase reactors** for methanation [12, 40–43]. Generally, a slurry reactor is filled with a liquid phase, in which fine catalyst particles are suspended as a result of the gas flow. The presence of the liquid phase with high heat capacity allows for effective and accurate temperature control: the heat of reaction can be completely removed and the reactor can operate almost isothermally, leading to a simple process design. The challenges with the use of slurry reactors are a result of gas liquid mass-transfer resistances and the decomposition and evaporation of the suspension liquid [11, 12, 40, 44].

Structured reactors such as monolith reactors are polytropic reactors. Due to their internal metallic structure, monolith reactors feature an enhancement of radial heat transport by two to three orders of magnitude due to heat conduction through the metallic structure (depending on the metallic material) [45]. A crucial advantage is the control of the temperature profile along the reactor, which can be specifically adjusted by process parameters (e.g. *GHSV*, cooling power). Hence, structured reactors can be kept at the level of the maximum reaction rate [46]. By this procedure falling into the range of the equilibrium limitations can be avoided (see Figure 3.1). Furthermore, the product gas leaves the reactor with moderate methanation temperatures (e.g. 300 °C) to enable high conversion rates with respect to thermodynamic limitations.

A special structured reactor concept is embodied by micro-structured reactors which are very compact reactors with a high surface-to-volume ratio. They combine high heat transfer and a small pressure drop [8, 47–49]. Drawbacks of structured reactors are the more complicated catalyst deposition on the metallic structure, as well as the difficulty of replacing the deactivated catalyst: once the catalyst has been deactivated, the whole reactor has to be equipped with a new catalyst coating.

A further development of structured reactors is the sorption enhanced methanation reaction concept. The water produced by the methanation reaction is removed from the gas phase by the catalyst carrier showing adsorbent functionality. Thereby, thermodynamic limitation of the conversion rate is reduced. For the subsequent water removal, temperature swing and/or pressure swing with or without purge gas can be applied. A further development of the Zurich University of Applied Sciences combines this water removal with a regeneration step where the applied materials can undergo catalytic and microstructure regeneration [50, 51].

3.3.2.1 Dynamic operation

Dynamic operation significantly changes the methanation catalyst and reactor requirements. Under dynamic operation, the reactor temperature can severely change if the reactor heating or cooling cannot be adapted quickly enough. This strong temperature variation can lead to catalyst cracking or sintering which diminishes the catalyst lifetime [2].

One way to overcome the temperature variation issue is to adapt methanation reactor concepts for dynamic temperature regulation. Rönsch *et al.* simulated the temperature increase of an adiabatic methanation reactor with outlet gas cooling and recirculation [52]. They showed that, even with an adaptive increase in gas recirculation, a catalyst temperature increase of more than 30 °C cannot be prevented after a strong increase in inlet gas flow. Isothermal reactor concepts may show better temperature regulation under dynamic conditions. Lefebvre *et al.* investigated the dynamic behaviour of a lab-scale three-phase methanation reactor and showed that the reactor temperature does not change even though the gas load at the reactor inlet was varied between 25 and 100% [12]. Consequently, the liquid phase makes the three-phase reactor less sensitive to fluctuating feed streams.

If the H_2 production from the electrolyser is not sufficient, the methanation reactor may have to be shutdown. In this case, a reactor with rapid start-up and shutdown is required. ZSW, Germany investigated the start-up and shutdown behaviour of a micro-methanation reactor with an oil cooling system [53]. They could not observe any catalyst deactivation but also reached high CO_2 conversion, which is a promising result for PtG technology.

3.3.2.2 Pilot and demonstration facilities

An overview of PtG demonstration plants is already available in literature [14, 22, 53]. Hereby, some of these pilot and demonstration plants are outlined.

The first PtG pilot plant was probably that which was built in Japan at the **Tohoku Institute of Technology** in 2003. The plant is relatively small, producing 1 m^3/h methane (electrical power at electrolyser <10 kW). Seawater is used for electrolysis [54].

The combination of PtG with biomass gasification is demonstrated in the **KIC-Project "DemoSNG"** [55]. A novel methanation concept based on honeycomb reactors is used. This concept was investigated in the last few years at the Engler-Bunte-Institute of KIT [56] and was transferred to demo scale in 2014. The integration of hydrogen from surplus electricity is optional with this reactor concept and offers increased total carbon exploitation (methanation of CO and CO_2) and process efficiency as additional value. Furthermore, the plant can be operated on partial load , which is especially useful for PtG operation linked to volatile electricity. This means a basic load for the methanation of CO from Syngas and additional the methanation of CO_2 from Syngas if hydrogen is produced by electrolysis. The methanation module has a methane production capacity of 100 kW.

The **BioPower2Gas** project aims to use the biological methanation developed by MicrobEnergy GmbH as part of the PtG process chain. A PEM electrolyser with an electrical power input of 400 kW provides the hydrogen. The plant, which is located in Allendorf, Germany, was recently put into operation [57].

Other demo projects such as the plants in Falkenhagen [58], Frankfurt [59] and Mainz [60] are related to the direct injection of H_2 into the gas grid.

3.4 Commercialisation, regulator and commercial outlook in Germany

3.4.1 Commercialisation in Germany

The world's only commercial PtG plant has been put into operation by Etogas GmbH in December 2013 for the car manufacturer Audi AG. This 6.3 MW plant produces H_2 from wind power and converts 2800 t CO_2 per year from a nearby biogas plant to supply fuel to approximately 1500 CNG cars. This represents a mileage of 15,000 km/a [61].

3.4.2 Regulation in Germany

Before injection into the German gas grid, SNG from a PtG plants needs to fulfil the DVGW G 262 standards: SNG needs to have a minimum CH_4 content of 95% (H-gas) or 90% (L-gas), respectively. If a natural gas filling station is present in the relevant gas grid, the hydrogen content is limited to a maximum of 2 vol.% in order to satisfy the DIN EN 51624 standards for public distributed fuels. These standards are summarised in Table 3.2.

3.4.3 Commercial outlook in Germany

Economic evaluation performed by Reference 53 showed that SNG from PtG processes is not competitive with natural gas or even biomethane produced via anaerobic fermentation in Germany ($\approx$7 ct/kWh). The SNG production costs from PtG calculated by different authors differ strongly [62]. However, it can be assumed that the PtG-SNG price is currently at least two times higher than for biomethane. For economic feasibility, different business cases such as mobility, balancing services and CO_2 certificates have to be combined [53]. With respect to a macroeconomic consideration, it has to be taken into account that PtG can contribute to minimise the expansion of the electricity grid infrastructure [63] and to increase the share of renewable energy in the transport and the heating sectors. Therefore, PtG can play a major role in the realisation of the ambitious transition of the energy system.

Table 3.2 Different standards for the injection of SNG in Germany (not all need to be fulfilled necessarily)

German standard	G 260	G 262 (H-gas)	G 262 (L-gas)	DIN EN 51624
Higher calorific value in kWh/m^3	8.4–13.1	–	–	–
CH_4 content in mol%	–(>76*)	>95	>90	–
H_2 content in vol.%	–	<5 (see CH_4)	<10	<2

*Equivalent to 8.4 kWh/m^3 (higher calorific value) if CH_4 is the only combustible gas.

3.5 Symbols and abbreviations

Abbreviations and indices

0	Standard conditions
3PM	Three-phase methanation
AEL	Alkaline electrolysis
DVGW	Deutscher Verein des Gas- und Wasserfaches
I	Component i
PEM	Polymer electrolyte membrane
PtG	Power-to-Gas
SNG	Substitute Natural Gas
SOEC	Solid Oxide Electrolysis
ZSW	Zentrum für Sonnenenergie- und Wasserstoff-Forschung Baden-Württemberg

Symbols

ΔH_r^0	Standard enthalpy of reaction	kJ/mol

References

[1] C. H. Bartholomew, "Carbon deposition in steam reforming and methanation," *Catalysis Reviews*, vol. 24, no. 1, pp. 67–112, 1982.

[2] C. H. Bartholomew, "Mechanism of catalyst deactivation," *Applied Catalysis A: General*, vol. 212, pp. 17–60, 2001.

[3] P. B. Tøttrup, "Kinetics of decomposition of carbon monoxide on a supported nickel catalyst," *Journal of Catalysis*, vol. 42, no. 1, pp. 29–36, http://www.sciencedirect.com/science/article/pii/0021951776900877, 1976.

[4] D. L. Trimm, "The formation and removal of coke from nickel catalyst," *Catalysis Reviews*, vol. 16, no. 1, pp. 155–189, http://dx.doi.org/10.1080/03602457708079636, 1977.

[5] J. Gao, Q. Liu, F. Gu, B. Liu, Z. Zhong, and F. Su, "Recent advances in methanation catalysts for the production of synthetic natural gas," *RSC Advances*, vol. 5, no. 29, pp. 22759–22776, http://dx.doi.org/10.1039/C4RA16114A, 2015.

[6] S. Fujita and N. Takezawa, "Difference in the selectivity of CO and CO_2 methanation reactions," *Chemical Engineering Journal*, vol. 68, no. 1, pp. 63–68, http://www.sciencedirect.com/science/article/pii/S1385894797000740, 1997.

[7] G. D. Weatherbee and C. H. Bartholomew, "Hydrogenation of CO_2 on group VIII metals: II. Kinetics and mechanism of CO_2 hydrogenation on nickel," *Journal of Catalysis*, vol. 77, no. 2, pp. 460–472, http://www.sciencedirect.com/science/article/pii/0021951782901865, 1982.

[8] K. Müller, M. Städter, F. Rachow, D. Hoffmannbeck, and D. Schmeißer, "Sabatier-based CO_2-methanation by catalytic conversion," *Environmental Earth Sciences*, vol. 70, no. 8, pp. 3771–3778, 2013.

[9] G. A. Mills and F. W. Steffgen, "Catalytic Methanation," *Catalysis Reviews*, vol. 8, no. 1, pp. 159–210, 1974.

[10] T. Inui, M. Funabiki, M. Suehiro, and T. Sezume, "Methanation of CO_2 and CO on supported nickel-based composite catalysts," *Journal of the Chemical Society, Faraday Transactions 1*, vol. 75, no. 0, pp. 787–802, http://dx.doi.org/10.1039/F19797500787, 1979.

[11] M. Götz, "Methanisierung im Dreiphasen-Reaktor," *Dissertation, Engler-Bunte-Institut, Bereich Chemische Energieträger – Brennstofftechnologie, Karlsruher Institut für Technologie (KIT)*, Karlsruhe, 2014.

[12] J. Lefebvre, M. Götz, S. Bajohr, R. Reimert, and T. Kolb, "Improvement of three-phase methanation reactor performance for steady-state and transient operation," *Fuel Processing Technology*, vol. 132, no. 0, pp. 83–90, http://www.sciencedirect.com/science/article/pii/S0378382014004640, 2015.

[13] W. Kreuter and H. Hofmann, "Electrolysis: the important energy transformer in a world of sustainable energy," *International Journal of Hydrogen Energy*, vol. 23, no. 8, pp. 661–666, http://www.sciencedirect.com/science/article/pii/S0360319997001092, 1998.

[14] G. Gahleitner, "Hydrogen from renewable electricity: an international review of power-to-gas pilot plants for stationary applications," *International Journal of Hydrogen Energy*, vol. 38, no. 5, pp. 2039–2061, 2013.

[15] Ø. Ulleberg, T. Nakken and A. Eté, "The wind/hydrogen demonstration system at Utsira in Norway: evaluation of system performance using operational data and updated hydrogen energy system modeling tools," *International Journal of Hydrogen Energy*, vol. 35, no. 5, pp. 1841–1852, http://www.sciencedirect.com/science/article/pii/S0360319909016759, 2010.

[16] T. Smolinka, M. Günther and J. Garche, "NOW-Studie – Stand und Entwicklungspotenzial der Wasserelektrolyse zur Herstellung von Wasserstoff aus regenerativen Energien," NOW-Project report, 2011.

[17] A. Ursua, L. M. Gandia and P. Sanchis, "Hydrogen production from water electrolysis: current status and future trends," *Proceedings of the IEEE*, vol. 100, no. 2, pp. 410–426, 2012.

[18] B. Hacker, P. Gesikiewicz and T. Smolinka, "Arbeitspaket 1b: systemoptimierung und Betriebsführung der PEM-Elektrolyse," *energie/wasser-praxis*, pp. 37–40, http://www.dvgw-innovation.de/fileadmin/dvgw/angebote/forschung/innovation/pdf/1411hacker.pdf, 2014.

[19] M. Carmo, D. L. Fritz, J. Mergel and D. Stolten, "A comprehensive review on PEM water electrolysis," *International Journal of Hydrogen Energy*, vol. 38, no. 12, pp. 4901–4934, http://www.sciencedirect.com/science/article/pii/S0360319913002607, 2013.

[20] R. Gazey, S. K. Salman and D. D. Aklil-D'Halluin, "A field application experience of integrating hydrogen technology with wind power in a remote island location," *Selected papers presented at the Ninth Grove Fuel Cell SymposiumNinth Grove Fuel Cell Symposium*, vol. 157, no. 2, pp. 841 847, http://www.sciencedirect.com/science/article/pii/S0378775305016526, 2006.
[21] G. Müller-Syring, M. Henel, W. Köppel, H. Mlaker, M. Sterner and T. Höcher, "Entwicklung von modularen Konzepten zur Erzeugung, Speicherung und Einspeisung von Wasserstoff und Methan ins Erdgasnetz," Abschlussbericht DVGW-Projekt G1-07-10, 2013.
[22] M. Sterner and I. Stadler, *Energiespeicher - Bedarf, Technologien, Integration*. Berlin: Springer Vieweg, 2014.
[23] Proton Onsite. Available: http://protononsite.com/products/m/#tab-overview, 2015.
[24] M. A. Laguna-Bercero, "Recent advances in high temperature electrolysis using solid oxide fuel cells: a review," *Journal of Power Sources*, vol. 203, pp. 4–16, http://www.sciencedirect.com/science/article/pii/S0378775311024384, 2012.
[25] A. Brisse, J. Schefold and M. Zahid, "High temperature water electrolysis in solid oxide cells," *International Journal of Hydrogen Energy*, vol. 33, no. 20, pp. 5375–5382, http://www.sciencedirect.com/science/article/pii/S0360319908008355, 2008.
[26] M. Pozzo, A. Lanzini and M. Santarelli, "Enhanced biomass-to-liquid (BTL) conversion process through high temperature co-electrolysis in a solid oxide electrolysis cell (SOEC)," *Fuel*, vol. 145, pp. 39–49, http://www.sciencedirect.com/science/article/pii/S0016236114012757, 2015.
[27] M. Landgraf,"Power to Gas: Storing the Wind and Sun in Natural Gas", *KIT Press release*, N° 44, http://www.helmeth.eu/images/joomlaplates/documents/KIT_PI_2014_044_engl_Power_to_Gas_Storing_the_Wind_and_Sun_in_Natural_Gas.pdf, 2014.
[28] De Saint Jean, Myriam, P. Baurens and C. Bouallou, "Parametric study of an efficient renewable power-to-substitute-natural-gas process including high-temperature steam electrolysis," *International Journal of Hydrogen Energy*, vol. 39, no. 30, pp. 17024–17039, http://www.sciencedirect.com/science/article/pii/S036031991402391X, 2014.
[29] R. Anghilante and J. Lefebvre, Eds, "Coupling of a 3 phase methanation reactor and a high temperature electrolyser using Matlab Simulink: European fuel cell forum AG," 2014.
[30] E. Giglio, A. Lanzini, M. Santarelli and P. Leone, "Synthetic natural gas via integrated high-temperature electrolysis and methanation: part I—energy performance," *Journal of Energy Storage*, vol. 1, pp. 22–37, http://www.sciencedirect.com/science/article/pii/S2352152X15000067, 2015.
[31] J. Kopyscinski, T. J. Schildhauer and S. M. A. Biollaz, "Production of synthetic natural gas (SNG) from coal and dry biomass – a technology review from 1950 to 2009," *Fuel*, vol. 89, no. 8, pp. 1763–1783, 2010.
[32] S. Rönsch and A. Ortwein, "Methanisierung von Synthesegasen – Grundlagen und Verfahrensentwicklungen," *Chemie Ingenieur Technik*, vol. 83, no. 8, pp. 1200–1208, 2011.

[33] T. Schaaf, J. Grünig, M. Schuster, T. Rothenfluh and A. Orth, "Methanation of CO_2 – storage of renewable energy in a gas distribution system," (English), *Energy, Sustainability and Society*, vol. 4, no. 1, pp. 1–14, http://dx.doi.org/10.1186/s13705-014-0029-1, 2014.
[34] EU, "General Annexes, Technology readiness levels (TRL)", *HORIZON 2020 – WORK PROGRAMME 2014–2015*, http://ec.europa.eu/research/participants/data/ref/h2020/wp/2014_2015/annexes/h2020-wp1415-annex-g-trl_en.pdf, 2014.
[35] T. Schaaf, J. Grünig, M. Schuster and A. Orth, "Speicherung von elektrischer Energie im Erdgasnetz – Methanisierung von CO_2-haltigen Gasen: storage of electrical energy in a gas distribution system – methanation of CO_2-containing gases," *Chemie Ingenieur Technik*, vol. 86, no. 4, pp. 476–485, http://dx.doi.org/10.1002/cite.201300144, 2014.
[36] i. kiendl, M. Klemm, A. Clemens and A. Herrman, "Dilute gas methanation of synthesis gas from biomass gasification," *Fuel*, vol. 123, pp. 211–217, http://www.sciencedirect.com/science/article/pii/S0016236114000465, 2014.
[37] M. Buxbaum, *Methanisierungsreaktor*: DE Patent App. DE201110121930, 2013.
[38] M. C. Seemann, T. J. Schildhauer and M. A. Biollaz Serge, "Fluidized bed methanation of wood-derived producer gas for the production of synthetic natural gas," *Industrial and Engineering Chemistry Res*, vol. 49, no. 15, pp. 7034–7038, http://dx.doi.org/10.1021/ie100510m, 2010.
[39] J. Kopyscinski, T. J. Schildhauer and S. M. Biollaz, "Methanation in a fluidized bed reactor with high initial CO partial pressure: part I—experimental investigation of hydrodynamics, mass transfer effects, and carbon deposition," *Chemical Engineering Science*, vol. 66, no. 5, pp. 924–934, 2011.
[40] M. Götz, S. Bajohr, F. Graf, R. Reimert and T. Kolb, "Einsatz eines Blasensäulenreaktors zur Methansynthese," *Chemie Ingenieur Technik*, vol. 85, no. 7, pp. 1–7, 2013.
[41] M. Götz, F. Ortloff, R. Reimert, O. Basha, B. I. Morsi and T. Kolb, "Evaluation of organic and ionic liquids for three-phase methanation and biogas purification processes," *Energy & Fuels*, vol. 27, no. 8, pp. 4705–4716, 2013.
[42] J. Zhang, Y. Bai, Q. Zhang, et al., "Low-temperature methanation of syngas in slurry phase over Zr-doped Ni/γ-Al2O3 catalysts prepared using different methods," *Fuel*, vol. 132, no. 0, pp. 211–218, http://www.sciencedirect.com/science/article/pii/S0016236114004281, 2014.
[43] F. Meng, Z. Li, J. Liu, X. Cui and H. Zheng, "Effect of promoter Ce on the structure and catalytic performance of Ni/Al_2O_3 catalyst for CO methanation in slurry-bed reactor," *Journal of Natural Gas Science and Engineering*, vol. 23, no. 0, pp. 250–258, http://www.sciencedirect.com/science/article/pii/S1875510015000542, 2015.
[44] M. Götz, R. Reimert, S. Bajohr, H. Schnetzer, J. Wimberg and J. S. Schubert Thomas, "Long-term thermal stability of selected ionic liquids in nitrogen and hydrogen atmosphere," *Thermochimica Acta*, vol. 600, no. 0, pp. 82–88, http://www.sciencedirect.com/science/article/pii/S0040603114005036, 2015.

[45] C. Janke, M. S. Duyar, M. Hoskins and R. Farrauto, "Catalytic and adsorption studies for the hydrogenation of CO_2 to methane," *Applied Catalysis B: Environmental*, vol. 152–153, no. 0, pp. 184–191, http://www.sciencedirect.com/science/article/pii/S0926337314000228, 2014.
[46] L. Kiewidt and J. Thöming, "Predicting optimal temperature profiles in single-stage fixed-bed reactors for CO_2-methanation," *Chemical Engineering Science*, vol. 132, no. 0, pp. 59–71, http://www.sciencedirect.com/science/article/pii/S0009250915002523, 2015.
[47] O. Görke, P. Pfeifer and K. Schubert, "Highly selective methanation by the use of a microchannel reactor," *Catalytic Microstructured Reactors Catalytic Microstructured Reactors*, vol. 110, no. 1–2, pp. 132–139, http://www.sciencedirect.com/science/article/pii/S0920586105006218, 2005.
[48] Z. Liu, B. Chu, X. Zhai, Y. Jin and Y. Cheng, "Total methanation of syngas to synthetic natural gas over Ni catalyst in a micro-channel reactor," *Fuel*, vol. 95, pp. 599–605, 2012.
[49] K. P. Brooks, J. Hu, H. Zhu and R. J. Kee, "Methanation of carbon dioxide by hydrogen reduction using the Sabatier process in microchannel reactors," *Chemical Engineering Science*, vol. 62, no. 4, pp. 1161–1170, 2007.
[50] A. Borgschulte, N. Gallandat, B. Probst, et al., "Sorption enhanced CO_2 methanation," *Physical Chemistry Chemical Physics*, vol. 15, no. 24, pp. 9620–9625, http://dx.doi.org/10.1039/C3CP51408K, 2013.
[51] S. Walspurger, G. D. Elzinga, J. W. Dijkstra, M. Sarić and W. G. Haije, "Sorption enhanced methanation for substitute natural gas production: experimental results and thermodynamic considerations," *Chemical Engineering Journal*, vol. 242, no. 0, pp. 379–386, http://www.sciencedirect.com/science/article/pii/S1385894713016264, 2014.
[52] S. Rönsch, S. Matthischke, M. Müller and P. Eichler, "Dynamische Simulation von Reaktoren zur Festbettmethanisierung: dynamic simulation of fixed-bed methanation reactors," *Chemie Ingenieur Technik*, vol. 86, no. 8, pp. 1198–1204, http://dx.doi.org/10.1002/cite.201300046, 2014.
[53] F. Graf, M. Götz, M. Henel, T. Schaaf and R. Tichler, "Techno-ökonomische Studie von Power-to-Gas-Konzepten," *DVGW external report*, http://www.dvgw-innovation.de/fileadmin/dvgw/angebote/forschung/innovation/pdf/g3_01_12_tp_b_d.pdf, 2014.
[54] K. Hashimoto, N. Kumagai, K. Izumiya, H. Takano and Z. Kato, "The production of renewable energy in the form of methane using electrolytic hydrogen generation," *Energy Sustainability and Society*, vol. 4, no. 1, p. 17, 2014.
[55] S. Bajohr, D. Schollenberger, D. Buchholz, T. Weinfurtner and M. Götz, "Kopplung der PtG-Technologie mit thermochemischer Biomassevergasung: Das KIC-Projekt 'DemoSNG'," *gwf – Gas/Erdgas*, pp. 470–475.
[56] M. P. Wolf, "Minimierung des Druckverlusts durch Optimierung der CO-Entfernungsstufe für ein stationäres PEM-Brennstoffzellenheizgerät," *PhD Thesis, Karlsruhe*, 2011.
[57] Viessmann Werke GmbH, *Power-to-Gas-Anlage in Betrieb genommen*, 2015.

[58] E.ON, *E.ON Power-to-Gas-Pilotanlage in Falkenhagen.*
[59] Mainova, *Power-to-Gas-Anlage speist Wasserstoff ins Gasverteilnetz.*
[60] Stadtwerke Mainz AG: *Energie Park Mainz.* Available: http://energiepark-mainz.de/ (2015, August 13).
[61] Etogas, *Industrielle 6,3 MW PtG-Anlage (Audi e-Gas-Anlage).* Available: http://www.etogas.com/de/referenzen/article///industrielle-63-mw-ptg-anlage-audi-e-gas-anlage/ (2015, August 13).
[62] M. Götz, J. Lefebvre, F. Mörs, et al., "Renewable Power-to-Gas: a technological and economic review," *Renewable Energy,* http://www.sciencedirect.com/science/article/pii/S0960148115301610, 2015.
[63] C. Baumann, K. Geschermann, A. Moser, et al., "Nutzen von Smart-Grid-Konzepten unter Berücksichtigung der Power-to-Gas-Technologie," *Energ / wasser-praxis.*

Chapter 4

Production of hydrogen from methane decarbonization into Power-to-Gas scenarios

Alberto Abánades[1]

4.1 Introduction

The implementation of a new energy system based on a substantial share of renewable primary energy sources constitutes a paramount scientific and practical challenge for our Society, and a must if environmental sustainability wants to be made compatible with a stable energy delivery for the social, industrial and economic development of humankind. In particular, the decoupling between the energy demand, either in location or in time, with the energy production dispatchable for most of the renewable technologies requires the development of innovative solutions in the field of energy storage and transport at large scale.

This fact may imply a deep transformation of the current energy system as a whole, making stronger the need for integration of every energy source with the transport network in such a way that its management weakness may be compensated to provide a reliable and stable satisfaction to the energy demand of the Society. In the current situation, every final energy vector (natural gas, petroleum products or electricity) are managed separately, without any significant interaction.

On one hand, natural gas and petroleum products, for instance, show no critical management problems as they can be easily stored (in fact, they are a sort of chemical form of energy storage), but they will be exhausted at any moment in the future and its utilization constitutes the main concern regarding global warning. On the other hand, electricity has become a fundamental energy carrier, with multiple inputs but with a very limited storage capability, except hydraulic pumping storage power plants. The development of distributed energy sources, smart grids or medium/high capacity batteries has been put in place to solve the management problem by adding flexibility and a limited storage capacity in a network in which

[1]Technical University of Madrid, Madrid, Spain

the contribution of intermittent energy inputs will become more and more important. In fact, all these actions remain into the philosophy of the current energy system, as they act in one of the vectors and keep the separated management between them, keeping their inherent advantages and disadvantages. A general approach to an energy system in which the integration of every primary energy source and end use is commonly buffering into the system may remove their current environmental and management handicaps.

One of those approaches is the Power-to-Gas (PtG) concept, in which the energy produced by an intermittent energy source is transform into a chemical specie that may serve as energy storage and/or transport media, taking advantage of the environmental sustainability of the source and the management of a storage capability. This chemical specie may be methane or hydrogen, and the intermittent energy is any renewable electricity source as, mainly, windmills and photovoltaics (PV) plants, which are the current technologies with lower production management capabilities. Additionally, any other electricity production technology may be involved as electrical biomass, concentrated solar thermal, hydraulic and even clean fossil plants under certain circumstances.

The choice of methane as a pivotal element in this concept is based on the fact that storage and transportation of methane are well established and their utilization technologies are mature in fields as electricity production, industrial and transportation. In that respect, the PtG system may have the management capabilities of current gas networks with a high storage capacity and the environmental benefits of using low CO_2 emission technologies.

Therefore, the PtG process consists on the conversion of surplus electricity produced by a low-carbon technology into a grid-compatible gas in two main steps: H_2 production by electrolysis and CH_4 or synthetic natural gas (SNG) production by methanation combining this H_2 with CO/CO_2 from an external source (e.g., carbon capture from a centralized source as a combined-cycle power plant). The schematic description of this concept is depicted in Figure 4.1. Some fraction of hydrogen (around 5–10%) may be transported and stored in combination with SNG as well, and used in power plants.

Wind and solar PV plants are considered in the energetic policy of many countries as the preferred technologies to achieve carbon emission reduction targets. There are technological, economic and geographical reasons for that choice as both technologies have arrived to certain degree of maturity and their economic viability does not depend so strongly on the latitude. Wind power may still be enhanced by the development of offshore parks and solar PV do not have the geographical constraints of concentrated solar thermal plants whose viability are currently limited to high irradiation zones (e.g., around the tropics). Nevertheless, wind and solar PV are poorly managed in a grid without the stability provided by base power plants (generally fossil) and a big share of the total production may present energy delivery instabilities and deficiencies that could led to temporal losses of electricity supply to the system.

In the PtG scheme, the electricity produced by those main renewable technologies will cover a safe share of the electricity demand taking into consideration

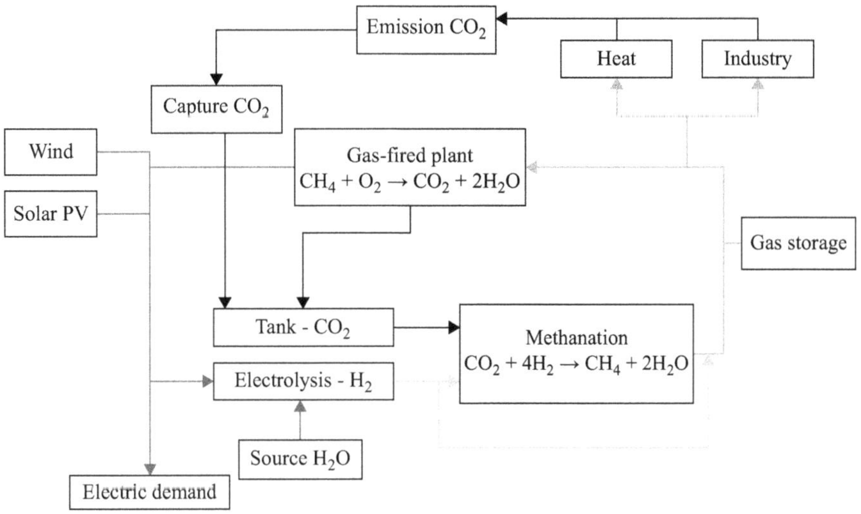

Figure 4.1 Basic scheme of the Power-to-Gas concept

its excess of capacity and foreseen availability, producing a substantial amount of surplus electricity that will be used to produce hydrogen via electrolysis.

The development of the PtG scheme has to face some critical issues:

- development and cost of hydrogen production technologies by electrolysis
- development of reliable methanation reactors
- development of efficient and environmentally friendly CO_2 capture technologies.

Into these schemes, carbon dioxide, hydrogen and methane are the key components that close the cycle. In particular, the CO_2 technology is one of the bottlenecks as it requires the development of CO_2 capture under various cases and its transportation. For the case of the CO_2 logistics, the technology may be considered mature, as liquid CO_2 transport trucks or CO_2 pipelines are available. The CO_2 capture is relatively mature for the case of flow streams with an important CO_2 concentration, as it is the case of coal fired and gas fired and coal gasification plants by mean of adsorption/absorption methods or membranes for its separation. Significant efforts are still required to reduce the operation, environmental costs and the energy penalty of these methods. Moreover, the development of carbon capture from other distributed sources (low scale and low CO_2 concentration) is on their first stages and is far from being available. This fact would make unlikely or economically unaffordable the recovery of CO_2 that could be produced by distributed thermal system in small industry and houses. This means that the CO_2 recovery may only be done in the medium term from thermal power plants that limits the CO_2 recycling rate as to effectively PtG be considered a close cycle in which enough CO_2 or carbon could be available to close entirely the cycle to produce the methane that is used as network storage. Therefore additional inputs to the system may be required to balance the cycle, especially from the carbon side, and fulfill the storage capability of the system.

In this chapter, the contribution of natural gas decarbonization will be discussed in its application to PtG.

4.2 Natural gas decarbonization

Natural gas decarbonization consists on the splitting in the absence of oxygen of the hydrocarbons that compound the natural gas into its components: solid carbon and hydrogen. Methane is the main component of the natural gas (and the pivotal gas into the PtG concept), and can be taken as the reference for the process that develops the following reaction:

$$CH_4 \rightarrow C + 2H_2 \quad (\Delta h_0 = 32.43\ \mathrm{kJ/mol}\ H_2)$$

This methane pyrolysis reaction is endothermic and must be driven by an external energy input that can be done in form of thermal or electric energy to brake the strong C–H bonds (dissociation energy: 436 kJ/mol). Such energy requirements implies that temperatures over 500 °C are needed to start the methane decomposition, reaching high conversion rates after 1 100 °C. Conversion rates from 10% at 500 °C (catalyzed by NiO [1]) to 95% at 1 050 °C (with carbonaceous catalyst [2]) has been reported.

The main characteristic of this process is the absent of oxygen, which produces null CO_2/CO impurities, making the process very suitable to produce hydrogen for any other chemical application, including the operation of fuel cells. On the other hand, it just converts methane into hydrogen and carbon, avoiding emissions, and substituting the management of CO_2 by carbon, constituting a technology that enables the utilization of fossil resources as methane in the context of a low-carbon economy. On the other hand, the full conversion into hydrogen is produced by one reaction. The final process has the potential to be more compact than other processes based on two or more reactions. For instance, no water-shift reaction is needed as in the case of methane steam reforming. In particular, its application to a PtG scheme introduces a complementary technology that can be applied in combination to a renewable energy input, stabilizing the system by the production of hydrogen in a managed way.

Methane decarbonization has a high potential of integration into the critical technological paths of the PtG networks. The direct and CO_2-free transformation of methane into hydrogen could be of great interest to make possible the utilization of SNG without CO_2 emissions. At this respect, end applications in industry, residential or mobility sector would not require carbon dioxide capture, which is one of the current main technological problems. The development of modular and competitive hydrogen production devices will reduce the need of local hydrogen storage capacity, which is a current technical challenge, and the need of hydrogen transport networks, as the system may run with existing natural gas transmission ducts, avoiding additional transport costs.

4.3 Status of the technology

The development of the methane decarbonization reaction has been explored by four different paths:

- plasma-torch driven methane pyrolysis
- direct thermal cracking at very high temperature
- catalyzed thermal decarbonization
- liquid metal driven methane decomposition.

The plasma-torch driven methane pyrolysis was already implemented industrially and similar methods based on the production of carbonaceous catalysts by plasma arc have been proposed. Among the concepts for this technique, it can be included the concept by Muradov *et al.* [3] or the Kvaerner CB&H process [4]. In this case, an industrial plant was built, producing 500 kg/h of pure carbon and 2 000 Nm^3/h of hydrogen, equivalent to 10 MW, with a consumption of 2.1 MW of electricity and 1 MW of high temperature steam. Unfortunately, most of the power input is provided by electricity and the size of the plant was not enough to demonstrate the economic viability of the technology for its application to industrial processes and distributed systems.

The direct thermal cracking is based on the thermal heating of methane up to high temperatures (above 1 300 °C) to arrive to the almost complete development of the methane cracking reaction. Originally, this technique has been proposed as technological mean to produce hydrogen for the production of clean solar fuels. The viability of this technique has been proven at laboratory scale, reporting practically complete methane to hydrogen conversions [5]. Nevertheless, very low hydrogen flow rates were obtained (of the order on mL/min), during limited period of times, with problems for a continuous operation due to the accumulation of carbon aggregates on the tubular structures in which the natural gas is flowing and heated to develop its decomposition. Some pilot-plant design for direct methane decomposition has been proposed for 10–50 kW-H_2 facilities.

Catalyzed thermal decarbonization arose as an alternative to develop the methane pyrolysis reaction with lower temperature constraints, which would imply an easier practical implementation of the technology. The identification of suitable catalysts has been widely studied for metallic and carbonaceous catalysts. Apart of the experimental test regarding the analysis of the heterogeneous catalysts and supports, the challenge is to design a continuous device that could be used at industrial scale. In brief, Nickel could be applied as catalyst for the reaction between 500 and 700 °C, which is a rather convenient temperature for an engineering industrial device. Nevertheless, as a consequence of the process, part of the carbon is deposited in the tip of the catalyst producing its deactivation.

The main drawback of catalyzed thermal decarbonization is the deactivation rate and the cost of the good catalysts. Reaching the economic viability of an industrial process based on this route requires the recycling of the catalyst in combination with a continuous operating reactor. The known techniques for catalyst regeneration rely on the air/steam oxidation of the encapsulating carbon that produced the catalyst deactivation, with the catalyst sintering risk and producing carbon dioxide and reducing its applicability to a low-carbon society. Different reactor designs have been successfully used alternating cracking and regeneration cycles in continuous mode, as parallel fixed-bed reactor and fluidized bed reactors. The most relevant implementation of catalyzed thermal decarbonization was the development and operation of the HYPRO process [6], based on fluidized bed reactors, with decomposition and regeneration sections, but it demonstrated high operation costs, unable to be competitive at that time.

Liquid metal driven methane decomposition has been recently proposed as a potential technology that could lead to the practical industrial implementation of methane cracking. Its main advantage is that it has not been reported critical clogging problems due to carbon accumulation, making possible the long-term operation of a reactor based on this technology [7, 8]. Liquid tin is proposed as molten media for the reaction. Additional work should be done to determine the scalability of this concept, but the high thermal diffusivity of liquid metals, and certain enhancement of the methane cracking reaction kinetics due to the catalytic effect of the metal or the carbon particles formed, are positive features in that sense.

Figure 4.2 shows the range of temperatures in which different methane decomposition technologies are proposed. The concepts based on the utilization of metallic catalyst achieve a reasonable performance at lower temperatures. Nickel-based catalysts are the most interesting up to date with operation between 500 and 700 °C. Other catalysts, such as iron-based materials, can compete for its lower cost, but require higher operation temperatures. Another alternative is the utilization of carbonaceous materials to enhance the decomposition reaction, such as carbon black, activated carbon or carbon nanotubes, operating at temperatures above 900 °C. The direct thermal decomposition has fast enough kinetics for the practical implementation of the reaction above 1 100 °C. Liquid metals, such as tin, enhance the heat transfer to the gas. Its catalytic performance is under discussion and research, although this approach with liquid metals as reaction medium suggests operation temperatures between 800 and 1 200 °C.

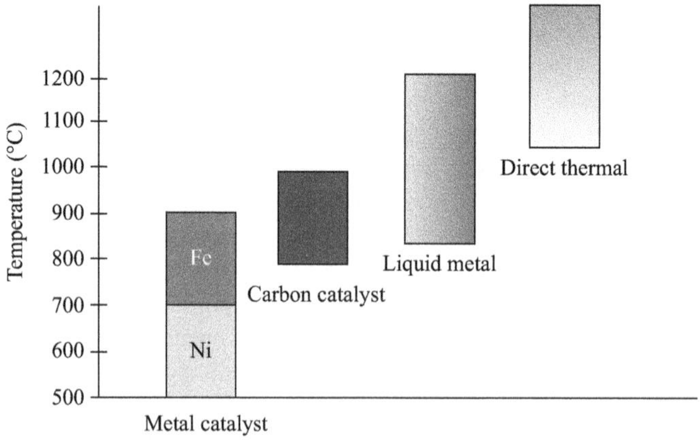

Figure 4.2 Temperature range application for catalyzed and direct thermal methane decomposition technologies

4.4 Introduction of methane decarbonization into PtG

The availability of technologies that enable the transformation between hydrogen and methane may be an important aspect of the PtG process. In that sense, methanation is the reaction that converts hydrogen into methane. The production of hydrogen is based on water electrolysis for the integration of renewables as wind and solar PV. Electrolysis may be an option for other electricity technologies that could help to develop the infrastructure for the system. Into those technologies, nuclear or thermal power plants could help to the system development during a transition phase, in which large capacity facilities should be built to satisfy the methane flow rates to be introduced in the system. Nevertheless, those technologies may be complemented by direct hydrogen production when using fossil resources. As an example of the efficient of the system for that purpose, the direct chemical decomposition of hydrocarbons has an efficiency if the conversion to hydrogen between 50% and 75%, corresponding to methane cracking and steam reforming [9]. In the case of using electricity from fossil origin, the efficiency of hydrogen production will be the efficiency of the thermal plant times the efficiency of the hydrogen production technology (41% in the best case with steam reforming and 58% gas combined cycle, including a few electric network losses by 5%).

Earlier in this section, the description of some possibilities to use methane decomposition into PtG is described. Its application has been considered upstream and downstream methanation stage. The impact and purpose is different on both two cases. The former is related to the production of hydrogen as material input for SNG production. The latter is applied to the possibility of hydrogen production and utilization on site that could be interesting for small scale facilities in the tertiary sector.

4.4.1 Methane decarbonization upstream PtG

The PtG scheme is originally based on the utilization of renewable electricity for the production of hydrogen that, in combination of recycle CO_2, serves for the production of synthetic methane. Renewable sources such as wind and PV will translate its intermittent production to the hydrogen feed of the methanation reactor. The compensation of this variability may be done by the consumption in the electrolyzer of network electricity produced, for instance, by gas fired plants. This approach seems to be clearly inefficient as it uses energy that it was previously obtained by methanation. The other alternative is the production of hydrogen following a manageable technology that could follow the variability of the renewable inputs as to keep stable the hydrogen feed to the methanation reactor. Methane decomposition may be used as a manageable hydrogen source for this purpose. Its integration on the PtG concept is shown in Figure 4.3.

The deployment of the PtG will be done gradually, in which the capacity and contribution of renewable electricity to the system will be increasing during a long period of time. In parallel, it should be developed the methanation step and the end-use technologies. In that sense, the possibility to have a low-carbon technology for the production of hydrogen decoupled the need to develop those three stages at the same pace. SNG facilities could be commissioned and operated independently of the availability of wind and PV to ensure a low CO_2 emission system. In that sense,

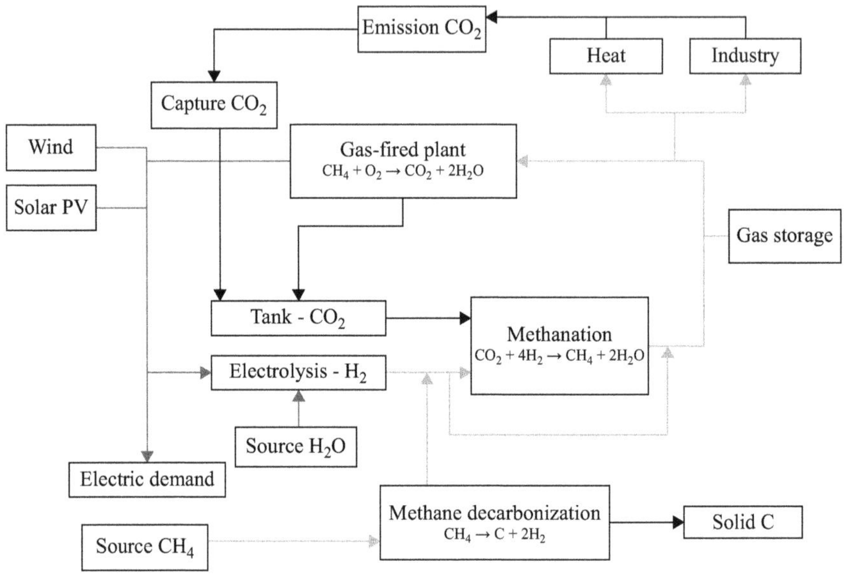

Figure 4.3 Methane decarbonization upstream the Power-to-Gas concept

methane decomposition can be considered a technology that will contribute to reduce emissions during the transition period between the starting of the operation of the PtG and its final implementation in the long term with the maximum share of renewable sources.

4.4.2 Methane decarbonization downstream PtG

The utilization of SNG as an energy carrier in a low-carbon energy system implies the need of technologies that could enable its utilization without CO_2 release. The options for these technologies are carbon dioxide capture or decarbonization.

The utilization of carbon dioxide capture implies the further application of CO_2 utilization technologies, or CO_2 sequestration. Methanation is one of the options in combination with the transportation of this CO_2 to the H_2 production sites to synthetize methane again. Nevertheless, this option requires a strong improvement of the capture technologies and its implementation at the CO_2 sources.

The development of methane decarbonization technologies that could be used from a few MW-H_2 (LHV) facility, corresponding to 150 kg/h, up to industrial process facilities producing up to 30 tons H_2/h, could be of great interest. The former corresponds to the size of a hydrogen refueling station. In that case, the application downstream the PtG scheme would allow the exploitation of the existing natural gas infrastructure to receive the methane and transform it to hydrogen on site eliminating the need for a hydrogen transport infrastructure.

As an example, a potential scenario takes into account the most populated areas in Europe as first sites for the development of the hydrogen economy up to

Figure 4.4 European hydrogen infrastructure in 2030 according to Reference 10

2030. Such scenario can be further extended by assuming 50% of the hydrogen to be provided by centralized systems and the other 50% by on-site production [10], following the structure shown in Figure 4.4. In previous forecasts of the automotive industry, it is foreseen a penetration of hydrogen vehicles between 7% and 22% of the market in 2030, with a demand between 2 and 7 Mton H_2 to feed between 12 and 40 millions of vehicles, with a 25% of such amount in Germany. These figures imply the establishment of approximately 10 000 fueling stations across Europe, of which 3 000 in Germany. For instance, the mentioned study [10]

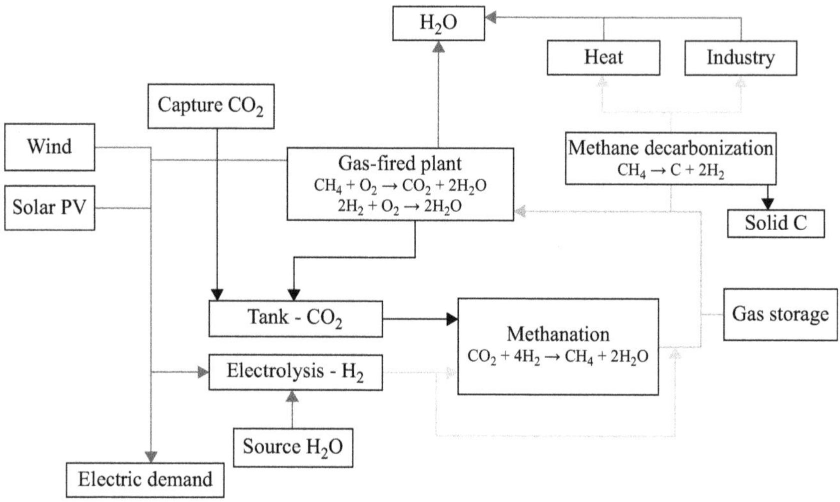

Figure 4.5 Integration of methane decomposition downstream into the Power-to-Gas concept

foresees a potential number of 900 stations in the Ruhr area, more than 600 in Berlin and 400 in Stuttgart in its most optimistic scenario.

Apart of the mobility case, the application of methane decarbonization may play an important role to make possible the carbon capture during the end use of methane in industries and houses. The development of small capacity units might make possible the transformation at small scale into the housing sector from methane to hydrogen, producing carbon as secondary material, that could be easily managed, and avoiding the risks of hydrogen storage or distribution to small consumers. A basic scheme for the integration of methane decomposition into PtG is depicted in Figure 4.5.

The application in large industries as hydrogen is at the base of several industrial processes. One of the most relevant ones is the production of Ammonia. This chemical is the starting feedstock for nitrogenous fertilizers synthesis.

In the case of housing systems, fuel cell of a production capacity of some kW-H_2 might be needed. Economic small scale hydrogen production systems from SNG without CO_2 emissions are needed, as the viability if the technology for CO_2 capture at that level will be likely difficult.

4.5 Socioeconomic impact of methane decarbonization in PtG

The PtG concept has as main leitmotiv the design of an energy system able to be managed with a high share of intermittent energy sources such as wind and PV, producing electricity that is managed by its substantial transformation into a chemical storage mean as hydrogen or methane.

The economic impact of this scheme depends on many factors, being a very important one the levelized cost of the energy contributors to that system. In particular, in the PtG scenario, in order to effectively reduce greenhouse gases emissions, the electricity input is provided mainly by solar PV plants and windmills.

For the case of wind-driven hydrogen production, the average cost is around 6 $/kg of hydrogen, that drops to the order of 3.80 $/kg for a centralized system [11]. For the case of solar PV, the cost estimations are higher, of the order of 9 $/kg [12], although its cost depends directly of the evolution of the electricity cost. Some estimations of the hydrogen production costs with methane decarbonization reaches 3 $/kg of hydrogen [13]. Therefore, low-carbon hydrogen produced by methane pyrolysis may be competitive integrated in comparison with the low emission alternatives. Nevertheless, the industrial viability of methane decarbonization may be confirmed.

From the economic point of view, the direct splitting of methane into carbon and hydrogen produces two potential assets, namely the hydrogen for its utilization in industry or as fuel, and the solid carbon that may have an economic worth, which can contribute to a lower cost of the final hydrogen product, being potentially competitive with current steam reforming, to cost as low as 1.5 $/kg of hydrogen.

The production cost of hydrogen in a large scale PtG scheme is a very important factor for the overall economics [14]. The deployment of the hydrogen economy will have another critical issue in the infrastructure that will be needed, for instance, for the penetration into the mobility sector to feed fuel cell vehicles. The competitiveness of hydrogen respect to petrol or gasoline is strongly handicapped by the lack of cheap hydrogen delivery infrastructures [14]. The application of methane decomposition processes that could be easily implemented into distributed hydrogen delivery centers may contribute to a simplification on the infrastructure required for its utilization by the end-users, as the current natural gas logistics may be used as energy transport and storage network.

4.6 Conclusion

In this chapter, the applicability of methane decarbonization in the PtG concept has been analyzed on a general sense. Some development is still needed to put into practice this technology, mainly to improve from the laboratory scale to the demonstration and industrial scale. Hydrogen production rates should be up-scaled from a few mL/min to some ton/h with economically viable processes that could compete with existing hydrogen production processes, either from methane (as steam reforming) or electrolysis.

The production of hydrogen from hydrocarbons without CO_2 emissions will have two possible routes, steam reforming with carbon capture and sequestration (CCS) and methane decomposition. Methane decomposition may be well integrated into PtG as it is a direct link between, components, and does not produce CO_2 in the process, which is the philosophy of the PtG'. This integration may be done before or after the methanation process.

The utilization of methane decomposition upstream methanation process enables the production of clean hydrogen during the transition to a high share of renewables in the electricity production system, could help to stability of the system compensating the intermittent production of sources such as wind or PV, and will decoupled the operation of the methanation stage from the electrolyzer.

Regarding the utilization of methane decomposition downstream methanation, it is a method to produce hydrogen from methane (the main storage material) with an effective capture of carbon, avoiding CO_2 emissions. This concept may be applied to big industrial facilities. Nevertheless, in high scale application, techniques for CO_2 capture may be competing and also viable. The application at small scale facilities for methane/hydrogen utilization could be very interesting in order to avoid CO_2 emissions that will be very unlikely captured and recycled.

A description of the integration of methane decarbonization in the PtG scheme is shown in Figure 4.6. The integration into the system will be affected by its economics. In that sense, methane decomposition could make available a clean hydrogen production technology at a cost competitive with steam reforming and electrolysis and could avoid the deployment of expensive hydrogen transport infrastructures.

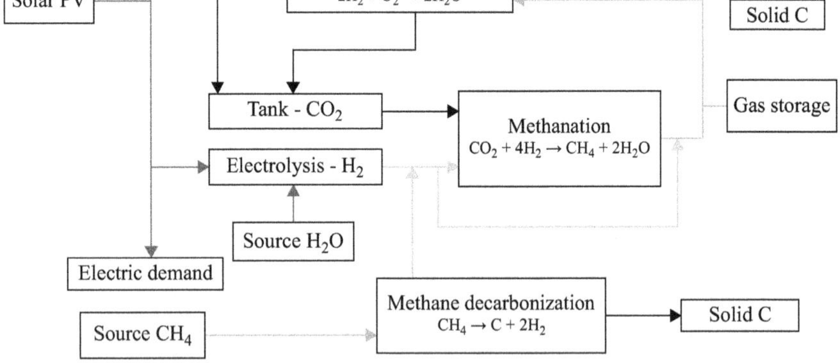

Figure 4.6 Basic scheme of the Power-to-Gas concept with the integration of methane decarbonization before and after methanation

References

[1] Y. Li, B.C. Zhang, X.W. Xie, J.L. Liu, Y.D. Xu, W.J. Shen. Novel Ni catalysts for methane decomposition to hydrogen and carbon nanofibers. *Journal of Catalysis* 238 (2006) 412–424.

[2] G. Maag, G. Zanganeh, A. Steinfeld. Solar thermal cracking of methane in a particle-flow reactor for the co-production of hydrogen and carbon. *International Journal of Hydrogen Energy* 34 (2009) 18 7676–7685.

[3] N. Muradov, F. Smith, G. Bockerman, K. Scammon. Thermocatalytic decomposition of natural gas over plasma-generated carbon aerosols for sustainable production of hydrogen and carbon. *Applied Catalysis A: General* 365 (2009) 292–300.

[4] B. Gaudernack, S. Lynum. Hydrogen from natural gas without release of CO_2 to the atmosphere. *International Journal of Hydrogen Energy* 23 (1998) 12 1087–1093.

[5] A. Abánades, E. Ruiz, E.M. Ferruelo, *et al.* Experimental analysis of direct thermal methane cracking. *International Journal of Hydrogen Energy* 36 (2011) 20 12877–1288.

[6] J. Poblenz, N. Scott. *Methods for hydrogen production by catalytic decomposition of a gaseous hydrocarbon stream*. 1966.

[7] A. Abanades, R.K. Rathnam, T. Geißler, *et al.* Development of methane decarbonisation based on liquid metal technology for CO_2-free production of hydrogen. *International Journal of Hydrogen Energy* 41 (2015) 8159–8167.

[8] T. Geißler, M. Plevan, A. Abanades, A. Heinzel, K. Mehravaran, R. Rathnam Experimental investigation and thermochemical modeling of methane pyrolysis in a liquid metal bubble column reactor with a packed bed. *International Journal of Hydrogen Energy* 40 (2015) (41) 14134–14146.

[9] M. Steinberg. Fossil fuel decarbonisation technology for mitigating global warning. *International Journal of Hydrogen Energy* 24 (1999) 8 771–777.

[10] E4tech (2005) *The Economics of a European Hydrogen Automotive Infrastructure*. Linde AG, http://www.hydrogenday.com/International/Web/Hydrogenday2005.nsf/e4tech%20hydrogen%20study.pdf

[11] U.S. Energy Information Administration. *The Impact of Increased Use of Hydrogen on Petroleum Consumption and Carbon Dioxide Emissions*. Report No. SR/OIAF-CNEAF/2008-04, August 2008, National Energy Information Center, EI 30, Washington, DC 20585. Available at: http://large.stanford.edu/courses/2010/ph240/usui1/docs/oiafcneaf(08)04.pdf

[12] T. Pregger, D. Graf, W. Krewitt, C. Sattler, M. Roeb, S. Möller. Prospects of solar thermal hydrogen production processes. *International Journal of Hydrogen Energy* 34 (2009) 10 4256–4267.

[13] A. Weimer, J. Dahl, J. Tamburini, A. Lewandowski, C. Bingham and F.C. Glatzmaier. Thermal dissociation of methane using a solar coupled

aerosol flow reactor. in *Proceedings of the 2000 DOE Hydrogen Programme Review,* NREL, San Ramon, California, USA.

[14] S. Shiebahn, T. Gribe, M. Robinius, V. Tiezte, B. Kumar, D. Stolten. Power to gas: technological overview, systems analysis and economic assessment for a case study in Germany. *International Journal of Hydrogen Energy* 40 (2015) 12 4285–4295.

Chapter 5

Power-to-Gas in an ancillary service market

Ushnik Mukherjee[1], Sean B. Walker[1] and Michael Fowler[1]

5.1 Introduction

Power-to-Gas (PtG) for energy storage is an energy hub concept that is gaining interest among many countries that are transitioning to an increasing amount of renewable energy in the power generation profile. The concept has been particularly attractive to countries with an existing natural gas (NG) system that can be used to store and distribute the electrolytic hydrogen produced. The PtG energy hub enables the power grid to utilize the volumetric capacity of the NG system for storing the off-peak surplus electricity or address the intermittent generation profiles of renewable energy systems. A review on existing PtG energy hubs that use renewable energy to produce hydrogen carried out by Gahleitner [1] in 2013 shows that there are a few PtG pilot plants located in Germany. A majority of these plants store the electrolytic hydrogen in hydrogen tanks and there are only a few pilot plants that make use of NG pipelines and underground gas storage reservoirs to distribute and store the gas over different temporal scales. Chapter 1 gives a detailed introduction to the concept and the technologies used within the PtG energy system framework.

This chapter explores the potential of the PtG energy hub to participate in the ancillary service market. The two most important functions of the power grid ancillary services management are to:

- balance generation output and load on the system on a continuous basis
- adjust output from generator and sometimes also dropping download by management of consumption at end users.

In order to operate the power grid in an effective manner, the operator procures different classes of ancillary services from both the suppliers and consumers of power. These services are categorized according to their response time, and the period over which they need to be offered for. Some of the most commonly known services include operating reserve services that have response times ranging from

[1]Department of Chemical Engineering, University of Waterloo, Waterloo, Ontario N2L 3G1, Canada

10 to 30 min and need to be offered for at least 2 h after a contingency event [2]. The Independent Electricity System Operator (IESO) in Ontario also procures ancillary services through demand response programs categorized as regulation service, a 5-min load following and hourly load following. The general goal of the demand response program is to incorporate end user facilities that can respond to price as well as load adjustment signals to vary their consumption over three temporal scales. Regulation service involves the grid operator sending the contracted end user to change their operating level on a second to second basis [3]. The 5-min and hourly load following on the other hand is a service procured from both generators as well as consumers of power. Understandably, the demand response load following program involves end users under a contract to curb their load during certain hours of the day by bidding into the market to offer load curtailment at a particular availability price [4]. In this chapter, a simple techno-economic feasibility analysis of the PtG energy hub to offer regulation service in the ancillary service market has been carried out.

In this case, the energy hub acts as a load to the grid, with the electrolyzer system being the major consumer. More importantly, the load also provide overall energy storage to the system by transforming electricity into hydrogen which can be stored and transported via the NG system to an energy demand. Since the electrolyzers can quickly ramp up and ramp down its energy consumption rate (or power levels), the system can respond to regulation signals to manage its consumption and at the same time be paid for the balancing service they provide to the grid. Participating in the ancillary service market gives the energy hub an additional revenue stream other than the ones from the three energy recovery pathways, namely the power to power, power to hydrogen (PtH), and the PtG energy recovery pathways described in Chapter 1.

According to the North American Electric Reliability Corporation (NERC) [5], the "power system" in North America is divided into four major interconnections which are described as "large machines" that work in a synchronized way to satisfy the electricity demand of its customers. Within each of these interconnections, there exists numerous balancing authorities that are responsible for meeting the energy demands for their respective customers by procuring services from generators that participate in the real time energy market.

5.2 Ancillary services: motivation

Due to the fluctuating nature of the load placed on the grid, it is difficult to match the generation output with it on a continuous basis at the local level, distribution level, and transmission grid level. The variation in load can be both short term as well as long term. Short term variability (minute-to-minute basis) in load is primarily attributable to the switching on and off of numerous individual loads in an instant. On the other hand, longer term variability (daily and seasonal) arises from factors like change in weather patterns over the course of a season which are generally foreseeable.

Fluctuations in voltage can also be observed on the generation side when generators experience variations in their scheduled output due to an unexpected failure in their equipments. The integration of intermittent renewables such as wind and solar generation units are also a cause of fluctuations observed on the generation side. This mismatch between supply and demand creates the fluctuations in frequency and voltage on the transmission and distribution lines of power. The disturbance in frequency caused due to the mismatch between supply and demand of power alters the number of oscillations of alternating current traveling from the supplier to the end user. Similarly, the variation in reactive power flowing between the supplier and the end user creates voltage fluctuations on the system [6].

Power system infrastructures throughout the globe try to procure ancillary services in order to maintain the reliability of the grid. Kirby [7] defines ancillary services as operations carried out by utilities that are responsible for producing, transmitting and controlling the flow of power on the network to assist in the effective operation of the entire system. The Alberta Electricity System Operator in their *Ancillary Services Participants Manual*, describes these services as ones that are used to operate the power grid in a way which enables the system to maintain the set standards for the operating frequency and voltage between the suppliers and end users of power [8].

According to Kirby, Rebours *et al.* and the NERC report [5–9], the term ancillary services includes different supporting provisions which are broadly classified into primary, secondary and tertiary control. Primary control is the first action taken after a contingency event that results in the interconnection frequency either falling or increasing to a level well outside the set dead band. It generally involves the use of both or either the generation units or responsive loads that maneuver their operating regimes respectively to arrest any further deviation in interconnection frequency. This fast action, taken within seconds after a frequency excursion event is one of the reasons why this control scheme is called frequency response. The control action is primarily automatic where generators sense an increase in the load (with the help of the governor) put on it when there is a change in the speed of the generator. The generator reacts to this decrease in speed by feeding more energy to its "prime movers" [5, 10]. As for the case of responsive loads, most of which are motors, they tend to reduce energy consumption and rotate at a lower velocity. It should be noted that primary control only helps in stabilizing small deviations in frequency. If the contingency event results in a significant drop in frequency which cannot be stabilized by primary control alone, secondary control is brought in to restore system reliability.

As a balancing authority has to meet both its internal needs while also satisfying its obligation toward helping in maintaining the interconnection frequency, there are bound to be instances when the scheduled inflow and outflow of power to and from the jurisdiction is different from the actual inflow and outflow. Thus this mismatch occurring while satisfying its internal and external obligations needs to be monitored. Most balancing authorities maintain a record of this by calculating the area control error (ACE) for their respective jurisdictions. The NERC report gives a detailed account on how the ACE is calculated for each balancing authority [5]. When the frequency is stabilized with the help of primary control after a contingency event in an interconnection, secondary control is employed to bring back the frequency to the level it was at before the contingency event. Secondary control employs the use of the ACE which is a value calculated in megawatts to decide how a collective effort of maneuvering generation capacity and load consumption will be able to bring the interconnection frequency back to its scheduled value (~60 Hz). One of the differences between primary and secondary control that is highlighted by UCTE [10] is that, primary control in general involves participation of all balancing authorities to maintain interconnection frequency, whereas secondary control is characteristic unique to every balancing authority, where they have to manage both the imbalances in generation and supply within the control area as well as tending to frequency balancing obligations from the interconnection as described earlier in this section. Secondary control is carried over a minute-to-minute time frame [5]. Tertiary control according to Rebours *et al.* [6] is a procedure employed when secondary frequency control cannot bring the interconnection frequency back to the scheduled value. It primarily involves altering the obligations of generation units, manually.

According to a presentation released by the IESO technical panel [11], the ancillary service market in the province procures regulation services through the use of existing pumped-hydroelectric power plants and reactive combined cycle gas turbine units. The regulation capacity procured is close to the 200 MW mark, but with the increase in the penetration of intermittent wind and solar generation units, even a 10% increase in their participation to the current grid infrastructure will result in a 50% increase in the regulation capacity requirement (300 MW).

In its earlier years, the ancillary services market had its rules and guidelines for offering the service setup around the operating technicalities of generators [12]. But due to the growth of the intermittent renewable energy portfolio in Ontario, the process of procuring ancillary services from generators has become even more complex. Although generators are the primary providers of ancillary services to the grid in Ontario, the IESO in recent years has also contracted "aggregated loads," and energy storage technologies such as flywheels and batteries to provide this service [13]. In 2015, Hydrogenics Inc. and Enbridge were recently contracted by the IESO to develop a 2 MW hydrogen production system that would offer regulation services in the greater Toronto area [14].

5.3 PtG energy hubs in the ancillary services market

The potential of electrolyzers to quickly ramp up and ramp down their operating levels enables the PtG concept to compete in the ancillary services market to offer regulation services. Work done by Eichman *et al.* [15] showed conclusive results in the maneuverability of electrolyzer operating levels. In their work, they assessed the total response time, ramp up and ramp down rates of a 40 kW polymer electrolyte membrane (PEM) and a 40 kW alkaline electrolyzer when subject to specific power use regulation signals. Tests for ramping down operating level by 25% for PEM electrolyzers showed that 99.1% of the required drop was completed within 0.2 s. As for a 75% ramp down in operating level, 96.7% of the required drop was completed within 0.2 s. Ramp down tests for both 25% and 75% in an alkaline electrolyzer were a bit slower but still in the high 90%. Ramp ups being more aggressive showed greater extent of completion in both systems. The alkaline electrolyzer, on average, completed 96.8% of the ramp up and 92.9% of the ramp down it was directed to reach within 0.2 s. Further analysis on the startup time for the PEM system showed that it could be turned on and ramped up to its rated capacity in a total time of 6 min and 27 s. The shutdown time analysis on the other hand was faster and showed that the electrolyzer can be ramped down and switched off from operating at its rated capacity in a total time of 1 min and 3 s, or just over a minute. It should be noted that larger scale units will have slower ramp rates, but still the values will be high enough to meet the response time requirements in dropping load for the different ancillary services. As for the duration for which the load regulation service needs to be provided, it is safe to assume that the electrolyzers can operate indefinitely for long periods of time. Although not covered in Eichman's work, the rate of degradation of the electrolyzer stack when subjected to frequent variations in operating level is one factor that needs to be assessed as well.

5.4 Multiservice PtG energy hub model

Figure 5.1 shows the overall PtG energy hub. It is of interest to see how a 2 MWh (nameplate capacity) electrolyzer system operates to offer regulation services. In order to assess the ability of the electrolyzer system to act as a load that can drop its energy consumption, it needs to have an incentive to run at levels other than what the regulation signal directs it to run at. Therefore, a pure hydrogen demand profile for 150 fuel cell vehicles (FCVs) is developed and added to the model so that there is a high value hydrogen demand. It is anticipated that FCV will be offered in the market in 2015–2020 by most large original equipment manufacturers [16]. Since the electrolyzer unit is located at a pressure reduction station, the hydrogen is injected and transmitted via distribution pipelines originating from that station. The pipelines bring the gas to a downstream station where with the help of pressure swing adsorption (PSA) units, pure H_2 is separated from NG and supplied to the FCVs. The model is also given the flexibility to offset some of the NG being sent to its end users when it is economically favorable.

Based on work done by Melaina *et al.* [17], it is assumed that the gas coming out of the pressure reduction station is at 29 bar and undergoes another cycle of pressure reduction to 1 bar at a downstream station. This pressure differential is considered to be suitable for PSA units to separate the hydrogen mixed in NG; the desired end use of the hydrogen is a pure stream of hydrogen removed from the electrolyzer. Note, that this is a benefit of the PtG concept in that the NG system can be used to transport hydrogen a low cost. For now in this work, it is assumed that there is no lag in the delivery of hydrogen to the end user.

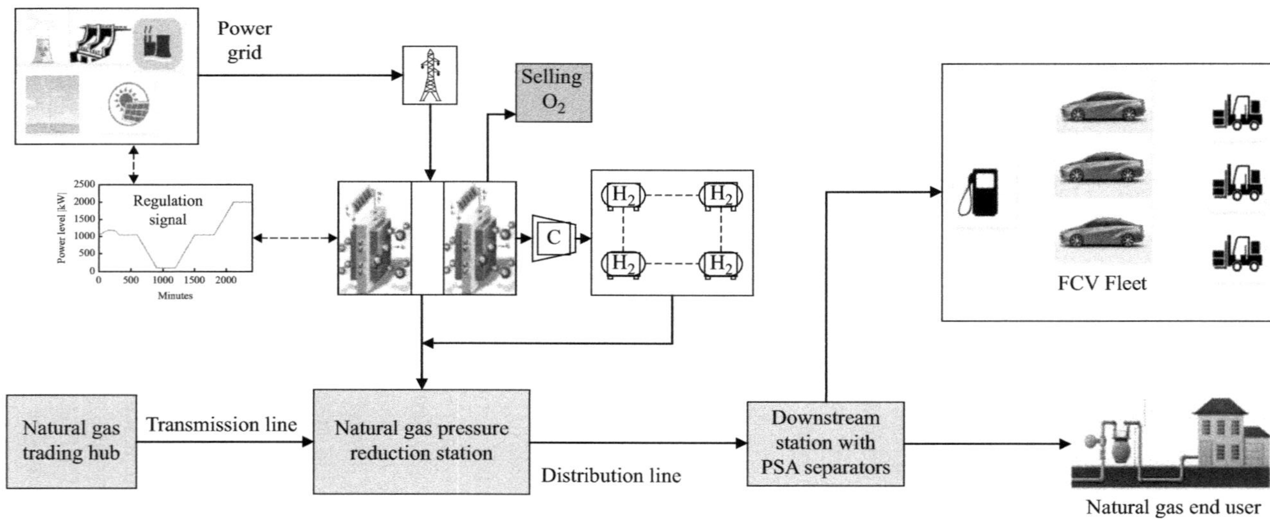

Figure 5.1 Overview of Power-to-Gas energy hub

150 FCVs refueling at 428.6 kg/day are chosen as an upper limit because that is the maximum number of cars the electrolyzer module can service. The demand profile for the vehicles is based on the 1st Annual University Student Design Contest: Hydrogen Fueling Station [18]. The variation in the demand over the week is based on a 2014 NREL report on Hydrogen Fueling Station in Honolulu, Hawaii which has a peak daily hydrogen demand of 65 kg [19].

5.5 Mixed integer linear programming problem formulation

A mixed integer linear programming (MILP) optimization problem has been formulated to assess a cost-effective way via which a PtG energy hub system can be operated to offer grid balancing services in Ontario while also benefiting from providing pure hydrogen and hydrogen enriched NG for the PtH and PtG energy recovery pathways.

The electrolyzer module used in the modeling study is a PEM electrolyzer developed by Hydrogenics Inc. To allow for a certain degree of flexibility, the model has the option to have a certain capacity of on-site compressed hydrogen storage. The types of tanks and compressors were chosen from a report released by Office of Energy Efficiency & Renewable Energy [20]. The optimization problem has the options to choose from two types of hydrogen storage tanks:

- Tank 1: maximum storage pressure: 430 bar, maximum capacity: 21.3 kg
- Tank 2: maximum storage pressure: 172 bar, maximum capacity: 89 kg.

To compress the gas to the required pressure, we also choose from four types of reciprocating compressors with each tank type having two types of compressors to pressurize the gas into it. The capacity of the reciprocating compressors vary from 42 to 251 kg/h. The efficiency for all the compressors according to IESO [20] is taken to be 65%.

The storage temperature for the gas is assumed to be 21 °C. This is the lower limit to which interstage coolers within a compressor unit cool the gas to, in order to reduce the power consumption during a compression cycle [20].

Figure 5.2 gives a systematic overview of how an optimization problem depicting the energy hub is formulated. The major inputs to the problem shown on the left hand side of the figure include parameters:

- Hourly Ontario Energy Price (HOEP) [21], which is the price paid for buying electricity to operate the electrolyzers;
- Natural gas (NG) spot price: This is the market price for NG, and the data is taken from the Henry Hub Spot Price data available from Reference 22. In this study, we sell hydrogen sold to the NG end user at the spot price;
- Due to the lack of regulation signal data available in the public domain of the IESO website, the signals used in this study has been taken from the Pennsylvania New Jersey Maryland (PJM) Interconnection's website [23]. The data set available on the PJM website has regulation signals listed every 2 s for

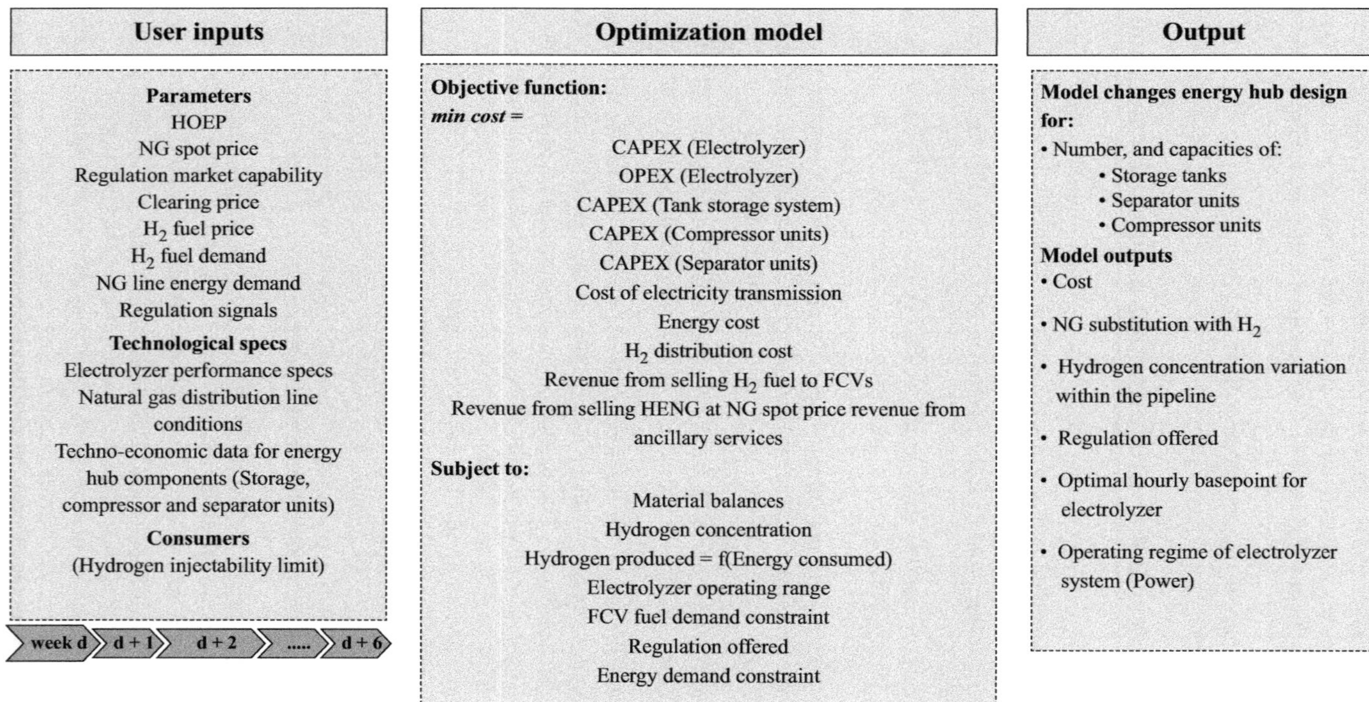

Figure 5.2 Formulation of the MILP optimization problem

a period of 40 min. In order to extend the data to an hour, the signals from the first 20 min has been repeated after the data ends at 40-min;

- The revenues earned from providing regulation services has also been based on the hourly regulation market capability clearing price set for the PJM jurisdiction [24];
- Other significant parameters include the hourly hydrogen demand (from 150 FCVs), a set pure hydrogen fuel price charged to the FCV owners [29], and technological specifications of the hydrogen production, storage, and compression systems.

The CPLEX algorithm in the general algebraic modeling system, version 22.4 has been used to solve the MILP problem. The problem uses two time indices, namely, "seconds" and "hours." The "seconds" index is used to account for the regulation signals being sent every 2 s, and the "hour" index accounts for the hourly hydrogen demand placed on the system. The problem has been limited to a total of 168 h (~a week), to avoid dealing with long solve time issues, as there are 1 800 signals sent to the electrolyzers every hour. For example, the operating power level of the electrolyzer can vary every 2 s over an hour, therefore, to arrive at the optimal value of the objective function, optimal power operating levels for the electrolyzer needs to be found for each of the 302 400 (168 × 1 800) s over the 168-h period.

The middle section in Figure 5.2 shows the objective function and lists some of the constraints that the MILP problem is subjected to. As mentioned previously, the objective of the study is to find the economic way of operating an energy hub of fixed hydrogen production module. Since cost minimization is the major objective, the capital cost and the operating and maintenance cost (2012, Canadian Dollars [CAD]) of the electrolyzer module is included in the objective function. The flexibility of having a buffer storage capacity, necessitates the inclusion of capital costs for the hydrogen storage and compression systems [20]. Since the system proposes to use the NG distribution line to distribute the hydrogen and extract it at a downstream station, the cost of installing separator units as well as the fuel cost to run compressors along NG pipeline [25] system needs to be accounted for. Note that the capital costs for the compression, storage, and separator units were modified to 2012 dollars (CAD) using the Chemical Engineering Plant Cost Index (CEPCI) [26]. Note that capital costs for all equipment includes their respective installation cost as well. An interest rate of 8% and a lifetime of 20 years have been assumed to amortize the capital cost of the components existing within the energy hub. The compressor and the electrolyzer system run on electricity; therefore, the cost associated with buying and transmitting electricity for the compressor and electrolysis has also been considered [27].

The optimization problem is subject to a set of equality and inequality constraints:

- The hydrogen demand constraint;
- Capacity offered by electrolyzer system for regulation. It should be noted that for this modeling study, the system is constrained to offer only 25% of the

actual modulation required by the regulation signal. This is done because a report released by the PJM website on Demand Side Response in Ancillary Service Markets specifies that demand resources are limited to providing 25% of the regulation requirement [28];

- Power operating limits for the electrolyzer module;
- Hydrogen concentration limit within the pipeline, which in this case is limited to 5 mol% (based on safeguards described in the work done by Melaina *et al.* [17]);
- Material balances over the pipeline and hydrogen storage system;
- Energy demand constraint placed by NG end users at the pressure reduction station. This constraint enables us to calculate the fraction of NG offset with hydrogen.

The MILP model is run under two different hydrogen pricing schemes and the results from the two studies are categorized under case study 1 and case study 2.

In case study 1, the pure hydrogen demand from the FCVs is sold at $5.08 per kg of hydrogen. This value is taken from a Department of Energy (DOE) report on hydrogen production cost from PEM electrolysis [29]. The hydrogen sent to the NG end user in this case is sold at the NG spot price on an energy content basis.

In case study 2, both the pure hydrogen as well as the hydrogen sent to the NG end user is sold at a premium. The selling price is the same as for pure hydrogen in case study 1 which is $5.08 per kg of hydrogen.

Figure 5.2 (right hand side) lists some of the desired outputs of the modeling study. The configuration of the energy hub, which primarily includes the size of the compression and storage system installed on site is one of the key results of the study. In addition to this, the model also gives the value of the cost objective, variations in pipeline hydrogen concentration, the amount of regulation offered every hour, and decides the optimal base point for the electrolyzer to operate at in each hour.

5.6 Results

This section elaborates and looks at how some of the parametric inputs to the MILP problem influences the desired outputs. It is seen that the configuration of the energy hub under the two different hydrogen pricing mechanisms remains the same. The model chooses one unit of tank type 1 which has a maximum storage capacity of 89 kg (172 bar: maximum storage pressure). The reciprocating compressor unit chosen for this tank is a system manufactured by RIX Industries that has a maximum capacity of 42 kg/h and can compress feed gas at 2.76–310.3 bar. In both cases, it is more economical to have one large tank rather than having multiple small tanks. Despite having the same configuration, the following paragraphs will talk about how the energy hub components are utilized in a different manner when the selling price of hydrogen is varied.

Figure 5.3 shows the variation in base point operating level (or power consumption level) of the electrolyzer (for case study 1). The MILP problem is

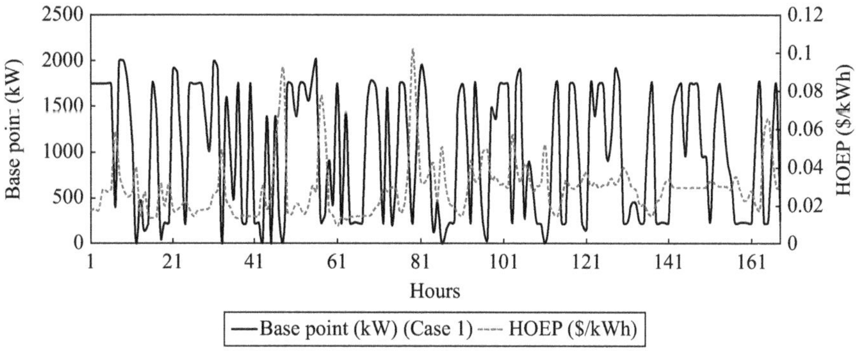

Figure 5.3 Case study 1: comparison of electrolyzer base point and energy price

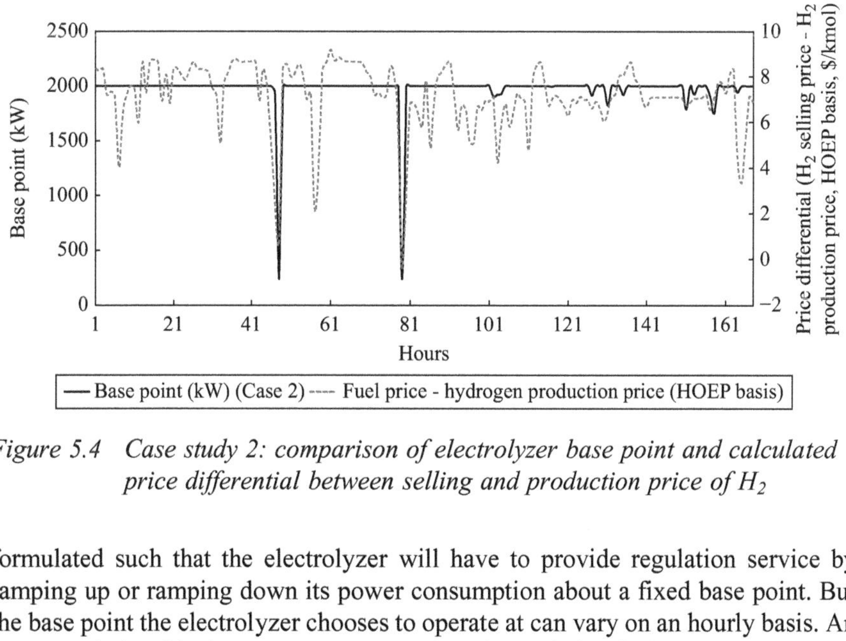

Figure 5.4 Case study 2: comparison of electrolyzer base point and calculated price differential between selling and production price of H_2

formulated such that the electrolyzer will have to provide regulation service by ramping up or ramping down its power consumption about a fixed base point. But the base point the electrolyzer chooses to operate at can vary on an hourly basis. An inverse relationship is seen upon comparing the variation in base point with the variation in HOEP. When the HOEP is high, the model chooses to operate at a lower base point operating level and offer regulation around it. So case 1 represents the case where energy arbitrage is a consideration.

In case study 2, there is no specific relationship between the HOEP and base point operating level of the electrolyzer. Since the selling price of hydrogen has been changed to the price set by the DOE [24], a difference between the hourly selling price and hourly production price of hydrogen was calculated. The hourly production price is calculated based on the efficiency of the electrolyzer and the energy it consumes to produce a kmol of H_2. This price differential has been plotted along with the hourly base point of the electrolyzer as seen in Figure 5.4. A drop in the price differential denotes a spike in the hydrogen production price. It is interesting to see that the base point in this case (Figure 5.4) decreases only when there is a significant spike in hydrogen production price.

As 1 800 regulation signals are sent to the electrolyzer over a period of an hour, the data were categorized into positive and negative signals. Since the signal changes every 2 s, the amount of variation in energy consumption while following a regulation signal can be calculated as follows:

$$\text{Regulation signal (kW)} \times 2\ \text{s} = \text{Energy modulation (kWs)} \tag{5.1}$$

To capture how much energy modulation is provided by the electrolyzer over an hour, the above equation can be modified to:

$$\text{Regulation signal (kW)} \times 2\ \text{s} \times \left(\frac{1\ \text{h}}{3\,600\ \text{s}}\right) = \text{Energy modulation (kWs)} \tag{5.2}$$

Using the same logic, when all of the negative regulation signals are summed up and divided by 1 800, the total reduction in energy consumption that the electrolyzer can provide over an hour can be calculated. The total increase in energy consumption while following all the positive regulation signals sent to the electrolyzer over an hour can be calculated in a similar way.

The total reduction in energy consumption that the system can offer is approximately 36.07 kWh. The red line in Figure 5.5 denotes the negative regulation offered by the electrolyzer system. In the hours where the system follows all of the negative regulation signals, the red line stays consistently close to the −36.07 kWh value (as shown by the horizontal portion of the red line). To analyze the parameters that influence when the system chooses to follow regulation, the normalized regulation market capability clearance price (RMCCP, Frequency

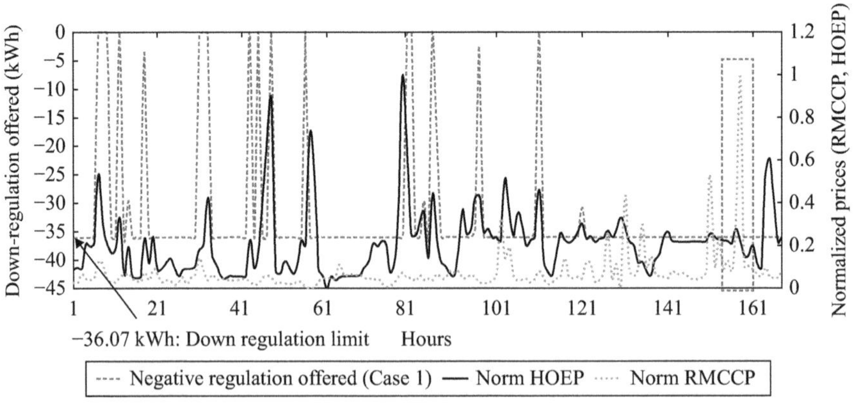

Figure 5.5 Case study 1: comparison of negative regulation offered by the electrolyzer system with the normalized regulation service price and the normalized energy price

Regulation Service) and the normalized HOEP were also plotted in Figure 5.5. The portion highlighted by red dotted lines on the right hand side of the figure shows that the system chooses to follow all of the negative regulation signals sent to it when the RMCCP is very high (as shown by the spike in the green line).

The pure hydrogen demand placed on the system also influences when the system chooses to follow the regulation signal and when it does not. The highlighted portion on the left hand side of Figure 5.6 shows that due to the high hydrogen demand and unfavorable RMCCP, the system chooses not to follow the regulation signal. In general, it is seen that whenever the RMCCP is greater than the HOEP, the system chooses reduces its energy consumption while also satisfying hydrogen demand. In order to satisfy both the regulation signal and the pure hydrogen demand, the system has the flexibility to either operate at a higher base point (when HOEP is significantly lower than RMCCP) or operate at a lower base point (when the RMCCP is not much greater than the HOEP) and withdraw gas from the storage tanks.

The total increase in energy consumption offered by the electrolyzer while following the positive regulation signals is approximately 66.62 kWh. The system is seen to follow all of the positive regulation signals in the hours in which the yellow line is horizontal and close to the 66.62 kWh mark. Similar to Figure 5.6, the portion highlighted by the red dotted line (in Figure 5.7) shows that the electrolyzer follows all of the regulation signal when the value associated with the normalized RMCCP is very high in comparison to the value of the normalized HOEP.

The portion highlighted by the black dotted line on the left hand side of Figure 5.8 shows that when the both HOEP is higher than the RMCCP, and the pure hydrogen demand is high, the system chooses to operate at the base point and not provide any regulation. On the other hand, the region highlighted on the right hand side (in Figure 5.8) shows that when the RMCCP is marginally greater than the

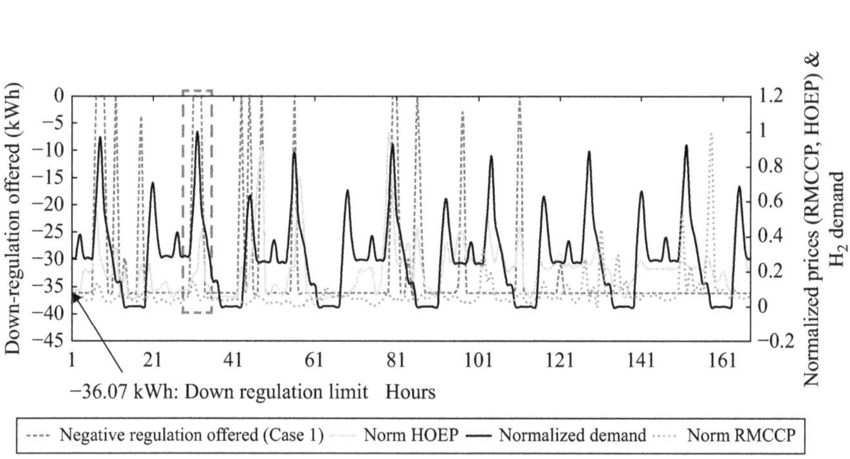

Figure 5.6 Case study 1: comparison of negative regulation offered by the electrolyzer system with the normalized regulation service price, normalized energy price, and the normalized pure hydrogen demand

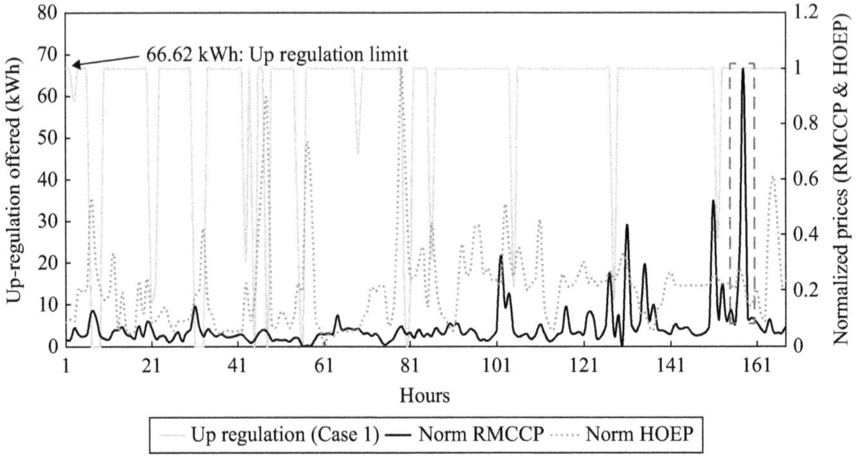

Figure 5.7 Case study 1: comparison of positive regulation offered by the electrolyzer system with normalized regulation service price and normalized energy price

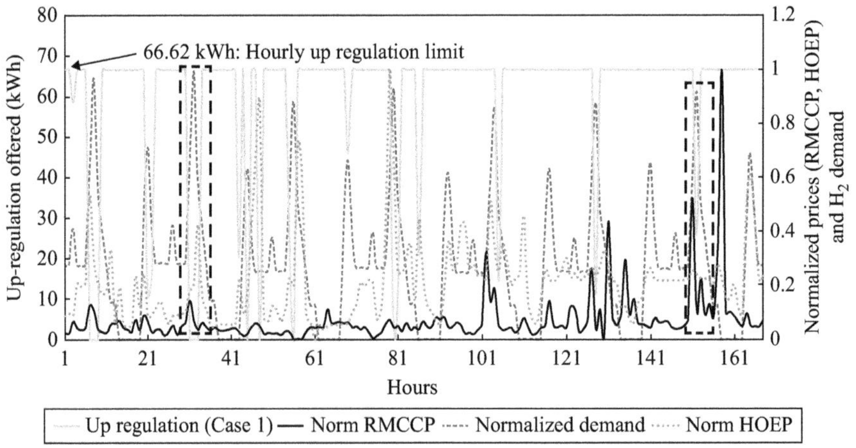

Figure 5.8 Case study 1: comparison of positive regulation offered by the electrolyzer system with the normalized regulation service price, normalized energy price, and the normalized pure hydrogen demand

HOEP, the system only follows some of the regulation signals (as indicated by the dip in the yellow line). As the pure hydrogen demand is high in this instance, the model tries to satisfy this peak in demand while maximizing its revenues by increasing consumption while following some of the positive regulation signals.

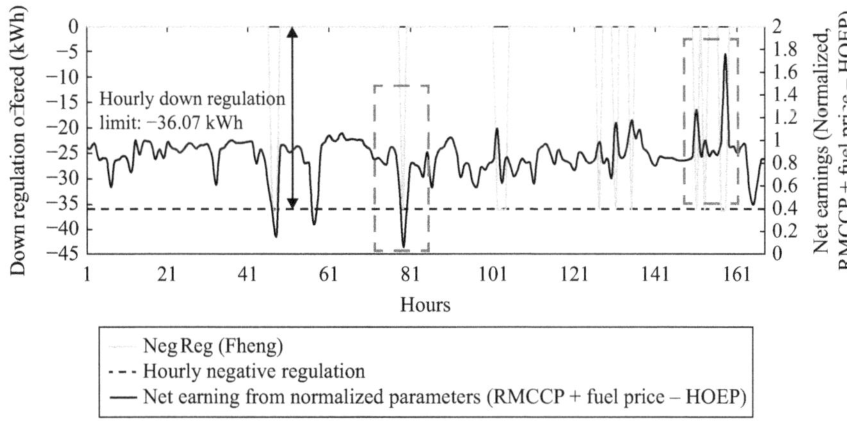

Figure 5.9 Case study 2: comparison of negative regulation offered by the electrolyzer system with the normalized net earnings which is the difference between the summation of normalized values of RMCCP and H_2 selling price (DOE) and the normalized HOEP

Figure 5.9 shows the negative regulation offered by the electrolyzer when hydrogen is sold at a premium price [24]. Similar to Figures 5.5 and 5.6, the total reduction in energy consumption that the system can offer over an hour while following the negative regulation signals is −36.07 kWh (shown by the dotted black horizontal line). The sharp troughs seen on the yellow trend line (in Figure 5.9) depicts the hours at which all of the negative regulation signals are followed. In this case, it is observed that the hours in which the system offers regulation is primarily influenced by the normalized value of net earnings in an hour. The term net earnings in this case implies the revenues the system would earn from providing regulation, and supplying H_2 to both FCV and NG end users while paying for the energy consumed to produce the hydrogen. Since the monetary values of the terms involved in calculating the net earnings are different, each of them were normalized. The region highlighted on the right hand side of Figure 5.9 gives an indication that the system provides regulation when the net earning is favorable. It is interesting to see how the system also tries to optimize the cost during the hours when net earnings are not favorable. The portion highlighted on the left hand side of Figure 5.9 shows that the system makes the most of the low net earnings by offering negative regulation. In this case, the HOEP is very high, but by reducing its energy consumption, the system can reduce the cost incurred while producing hydrogen and at the same time benefit from earning credits for the regulation provided.

The positive regulation offered by the system in case study 2 (as seen in Figure 5.10) is also influenced by the potential net earnings. The sharp spikes in the orange trend line (denotes amount of positive regulation provided) which coincide with the spikes in the brown line (denotes hourly net earnings) show that when the net earnings increase, the system follows the positive regulation signals

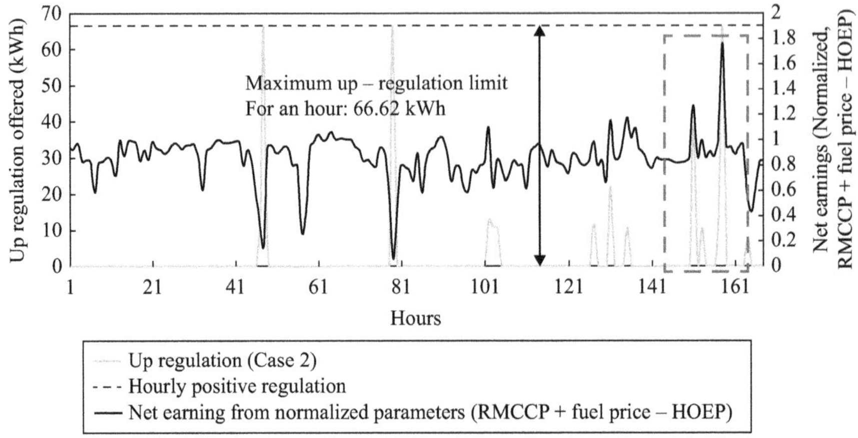

Figure 5.10 Case study 2: comparison of positive regulation offered by the electrolyzer system with the normalized net earnings which is the difference between the summation of normalized values of RMCCP and H_2 selling price (DOE) and the normalized HOEP

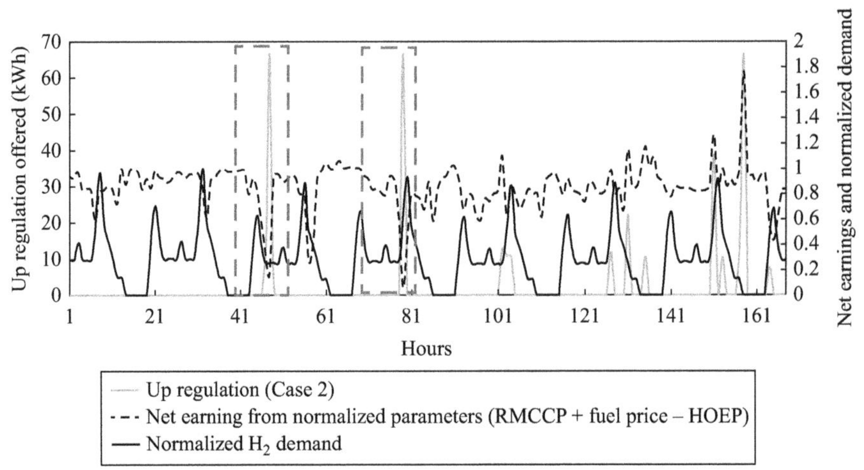

Figure 5.11 Case study 2: comparison of positive regulation offered by the electrolyzer system with the normalized net earnings which is the difference between the summation of normalized values of RMCCP and H_2 selling price (DOE) and the normalized HOEP, and also with the normalized H_2 demand

sent to it. Although this holds true for some instances, it is seen that the system provides positive regulation even when there is a drop in the net earnings.

In order to assess why the model chose to follow the positive regulation signals, the normalized hydrogen demand at those instances has been compared with the positive regulation trend line as seen in Figure 5.11.

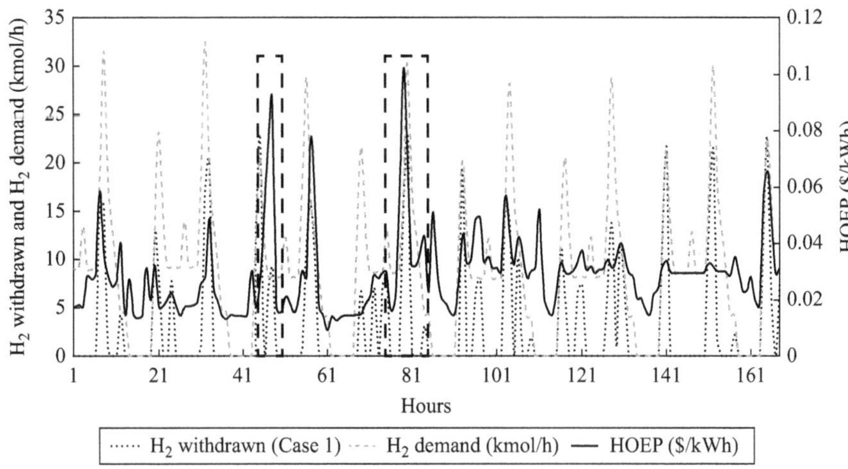

Figure 5.12 Case study 1: analysis on the influence of H_2 demand and HOEP on the withdrawal of the H_2 from storage tank

The highlighted regions in Figure 5.11 show the hours during which H_2 demand is greater than net earnings. Due to this fact, the model chooses to provide positive regulation around a lower base point (refer to Figure 5.4) that way the system makes the increase in energy consumption a more economical decision.

Figure 5.12 shows the periods during which hydrogen is withdrawn from the storage tank in case study 1. In order to minimize the cost of producing hydrogen during periods of high HOEP and high hydrogen demand, the system chooses to withdraw the gas from the storage tank as seen in the regions highlighted by the dotted lines.

A similar trend is seen in case study 2 where spikes in both the hydrogen demand and the HOEP forces the system to curb the cost for producing extra hydrogen by withdrawing gas from storage. Hydrogen is also withdrawn in periods when the HOEP is moderate because the hydrogen demand is seen to spike during those instances (Figure 5.13).

Figure 5.14 is a plot that shows the periods during which the model decides that it is beneficial to produce extra hydrogen and store some of it in the storage tanks. When compared with both the variation in hydrogen demand and HOEP, the system chooses to inject hydrogen in instances where there is a dip in both demand and energy price.

In case study 2 (Figure 5.15), the frequency of injection is lower than what is seen in case study 1 (Figure 5.14). The highlighted regions (in Figure 5.15) show that hydrogen is generally injected in periods when there is no hydrogen demand along with a dip in the HOEP. In other words, having a buffer storage capacity in the energy hub enables it to achieve energy arbitrage where it produces and stores extra hydrogen during periods of low energy price and low hydrogen demand so

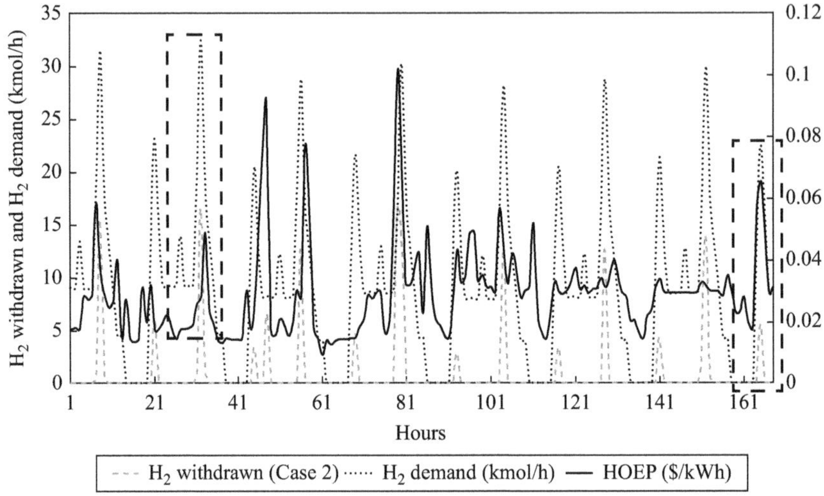

Figure 5.13 Case study 2: analysis on the influence of H_2 demand and HOEP on the withdrawal of H_2 from storage tank

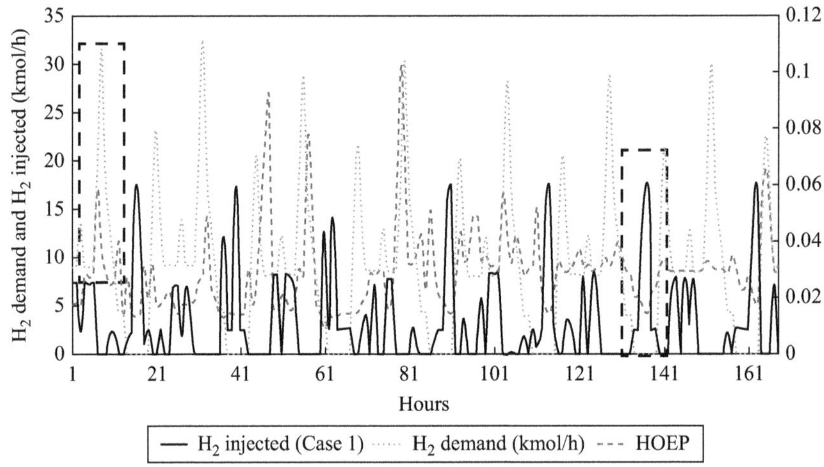

Figure 5.14 Case study 1: analysis on the influence of H_2 demand and HOEP on the injection of H_2 in to the storage tank

that it can withdraw the gas in periods of high energy price and high demand. In this way, the model minimizes the value of the cost objective function.

In order to see the effect of varying the hydrogen selling price to NG end users, the concentration of hydrogen within the NG pipeline pre-separation and post-separation for both case studies is shown in Figure 5.16. As case study 2 sets a premium price for hydrogen, the model has an incentive to produce extra hydrogen

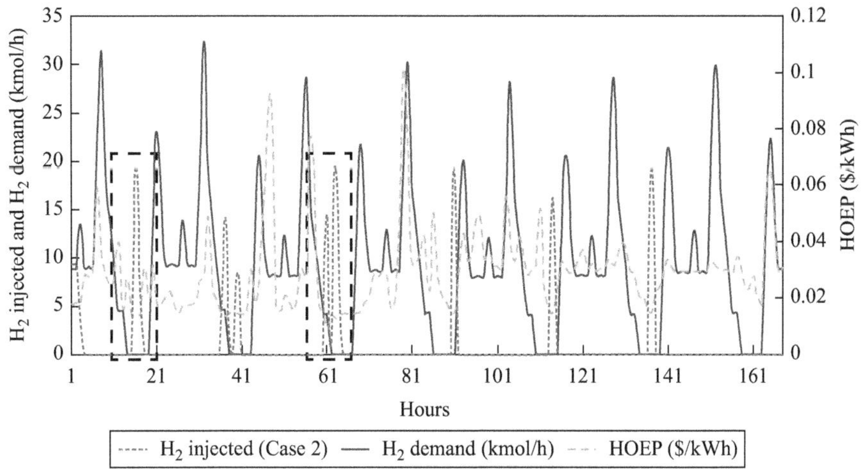

Figure 5.15 Case study 2: analysis on the influence of H_2 demand and HOEP on the injection of H_2 in to the storage tank

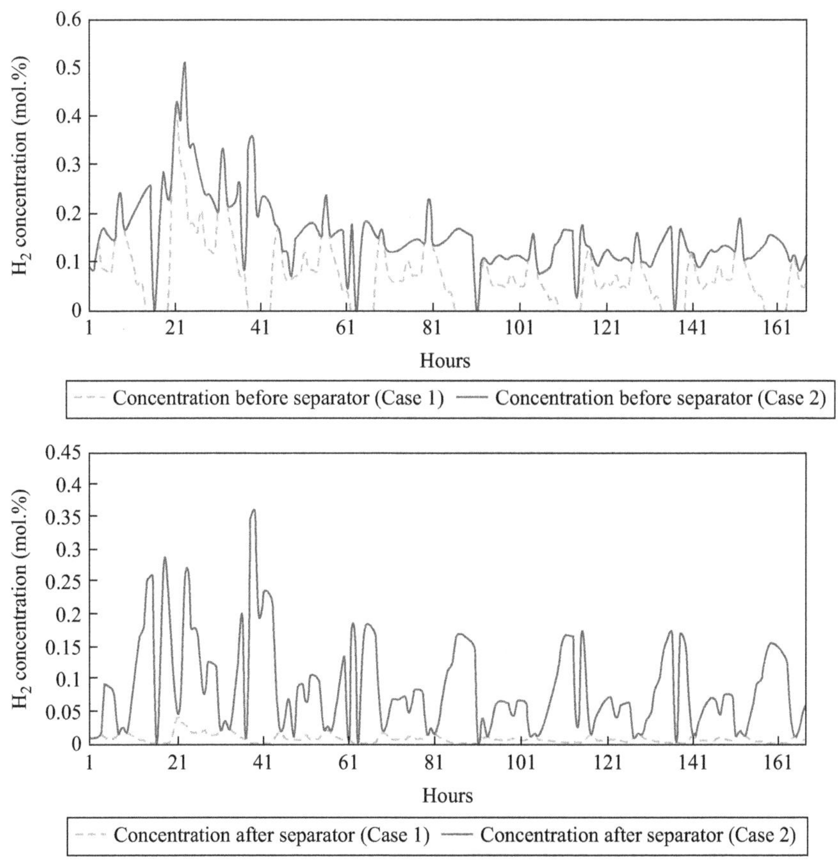

Figure 5.16 Case studies 1 and 2: hydrogen concentration variation within the natural gas pipeline both pre-separation and post-separation

and sell it to the NG end user. Therefore, the concentration of hydrogen is seen to be higher both before and after a certain amount of the gas is separated to satisfy the pure hydrogen demand. In case study 1, as hydrogen is sold at the NG spot price, the system does not produce any excess hydrogen to sell to the NG end user. The hydrogen that is not being separated due to the efficiency of the separator being less than unity (~90%), is the actual amount of hydrogen that is sent to the NG end user by default.

Although the model was run for a week, extrapolating the revenues to a year gives us a sense of the payback period for the project. The total capital for the system is same for both cases and is equal to $14 065 129. Case study 1 has a payback period of 17 years, where pure hydrogen is sold at a premium price to FCVs and at the NG spot price to NG end users. Case study 2 has a shorter payback period of 8 years, where H_2 is sold at a premium to both the transportation sector and the NG end user.

5.7 Conclusions

The results presented in Section 5.6 give an insight into the parameters that influence the decision variables in the optimization problem. The MILP formulation enables the electrolyzers to choose a base point (power operating) level every hour about which it chooses to vary its consumption. A parametric sensitivity analysis in the two case studies show that the HOEP as well as the selling price of hydrogen injected into to the NG grid influences how the elecrolyzer system operates.

In case study 1, the electrolyzer system is producing hydrogen that is sold to the FCVs at a premium price and at the NG spot price to the NG end user. The hourly base point of the electrolyzer system in this case has an inverse relationship with the HOEP, and is seen to decrease when the HOEP increases. Most of the regulation signals sent to the electrolyzer in this case is followed. A comparison between the RMCCP and the electricity price shows that in periods when the RMCCP is greater than the electricity price, the system follows all of the signals sent to it. Regulation signals sent to the electrolyzer (in case study 1) are not followed in periods where the service price is less than the electricity price. In some hours, it is seen that the system chooses to only follow the positive regulation signals. This is primarily attributed to the spike in hydrogen demand in those hours and the system tries to optimize the cost objective by earning revenues through positive regulation while meeting a surge in hydrogen demand.

As hydrogen is sold at the premium price to both the transportation as well as the NG end user in case study 2, the base point operating level of the electrolyzer for most of the simulation time period remains at the maximum nameplate capacity of 2 MW. The operating level is only seen to drop in case there is a significant difference between the hydrogen selling and hydrogen production price (Figure 5.4). In these hours when the base point drops, the system has the flexibility to follow the ramping up and ramping down signals sent to it. As a result of the electrolyzer running close to its maximum capacity, the system optimizes the cost objective by earning revenues while following the ramp down signals in hours of unfavorable net earnings caused by a spike in the electricity price (HOEP). In hours when the selling price of hydrogen contributes to a higher share of net earnings in comparison to revenues from regulation service, the system chooses to run at the maximum operating limit. Therefore, it is easier for the system to provide down regulation.

Due to a rise in the regulation service price, its share in the net earnings is higher than the share of hydrogen selling price during the latter hours of the week (~145–165). As a result of this, the electrolyzer system chooses to lower its base point operating level from the maximum capacity. Lowering the base point enables the system to ramp up and ramp down around the chosen operating level for those specific hours when the net earnings spike.

There also exists hours during which the net earnings (in case study 2) are not very high, but the system still chooses to follow the positive regulation signals. This is mainly because of the fact that there is an increase in the hydrogen demand. The economic incentive from running at the maximum operating limit and satisfying the peak in hydrogen demand is less optimal when compared to lowering the base point (operating level), and use the positive regulation signals to increase its production while meeting the increase in hydrogen demand.

The hydrogen concentration in the distribution pipelines post-separation (for case study 1) is the hydrogen that could not be separated from the incoming hydrogen enriched NG stream by the PSA unit due to its efficiency constraint. However in case study 2, the model chooses to produce more hydrogen and sell it to the NG end user because of the higher selling price of hydrogen. This highlights that it is not cost effective to sell hydrogen at the NG spot price. Although not assessed in this work, if one takes into account the emission reduction credits that can be earned when hydrogen enriched NG is sent to the NG end user, then selling hydrogen at the NG spot price can be a viable option. The potential of the PtG energy hub to provide multiple services like PtH for the transportation sector, PtG for the NG end user, and regulation services to the grid will enable the energy hub to have a faster payback period. The potential benefits of having systems like this is that it helps in initiating the shift to a H_2 economy while enabling the grid to achieve a better system operability. Although not included in this study, the emission reduction credits earned by the system when we use hydrogen as a fuel in the transportation sector as well will enable it to earn greater revenue. In order to fully realize the benefits of this concept while still being economical, the pricing schemes for hydrogen as well as for the regulation service provided need to be refined further.

References

[1] Gahleitner, G. (2013). Hydrogen from renewable electricity: an international review of power-to-gas pilot plants for stationary applications. *International Journal of Hydrogen Energy*, 38(5), 2039–2061.

[2] Operating Reserve Markets – Independent Electricity System Operator. (2015). Retrieved 5 June, 2015 from http://www.ieso.ca/Pages/Participate/Markets-and-Programs/Operating-Reserve-Markets.aspx
[3] Demand Response: A New Form of Regulation Service (IESO). (2015). Retrieved 7 June, 2015 from http://www.ieso.ca/Pages/Ontario's-Power-System/Reliability-Through-Markets/DR%20%26%20Regulation%20Service.aspx
[4] Demand Response Pilot: IESO. (2015). Retrieved 9 June, 2015 from http://www.ieso.ca/Pages/participate/Demand-Response-Pilot/default.aspx
[5] Balancing and frequency control: A technical document. (2011). North American Electric Reliability Corporation.
[6] Rebours, Y. G., Kirschen, D. S., Trotignon, M., and Rossignol, S. (2007). A survey of frequency and voltage control ancillary services—part I: technical features. *IEEE Transactions on Power Systems*, 22(1), 350–357.
[7] Kirby, B. J. (2004). *Frequency Regulation Basics and Trends.* (No. ORNL/TM-2004/291). Office of Scientific and Technical Information.
[8] *Ancillary services participant manual* (2010). (Operating Manual). Calgary, Alberta: Alberta Electricity System Operator.
[9] Rebours, Y. G., Kirschen, D. S., Trotignon, M., and Rossignol, S. (2007). A survey of frequency and voltage control ancillary services—part II: economic features. *IEEE Transactions on Power Systems*, 22(1), 358–366.
[10] Balancing energy: E-control. (2015). Retrieved 2 June, 2015 from http://www.e-control.at/businesses/electricity/electricity-market/balancing-energy
[11] *Proposed market rule changes – ancillary service contract terms.* (2012). (No. IESO TP 257-3b). IESO Technical Panel.
[12] Kirby, B. J. (2006). *Demand response for power system reliability: FAQ.* (No. ORNL/TM-2006/565). Tennessee: Oak Ridge National Laboratory.
[13] Procurement market: independent electricity system operator. (2014). Retrieved 27 October, 2014 from http://www.ieso.ca/Pages/Participate/Markets-and-Programs/Procurement-Market.aspx
[14] John, J. S. (2014). Ontario to become energy lab with 34 MW of 'all-of-the-above' energy storage. Retrieved 26 October, 2014 from https://www.greentechmedia.com/articles/read/ontario-contracts-34mw-of-all-of-the-above-energy-storage
[15] Eichman, J., Harrison, K., and Peters, M. (2014). *Novel electrolyzer applications: providing more than just hydrogen.* (No. NREL/TP-5400–61758). Golden, Colorado: NREL.
[16] Toyota to sell fuel-cell mirai worldwide in 2015. (2014). Retrieved 16 December, 2014 from http://www.ctvnews.ca/autos/toyota-to-sell-fuel-cell-mirai-worldwide-in-2015–1.2107063
[17] Melaina, M., Antonia, O., and Penev, M. (2013). *Blending hydrogen into natural gas pipeline networks: A review of key issues.* (No. NREL/TP-5600–51995). Golden, Colorado: NREL.
[18] *1st Annual University Student Design Contest: Hydrogen Fueling Station.* (2004). University of Waterloo, Waterloo, Ontario.

[19] Hill, P., and Penev, M. (2014). *Hydrogen fueling station in Honolulu, Hawaii feasibility analysis*. (No. INL/EXT-14–31624). Office of Energy Efficiency & Renewable Energy, Idaho National Laboratory, NREL.
[20] Nexant, Air Liquide, Argonne National Laboratory, Chevron Technology, Gas Technology Institute, National Renewable Energy Laboratory, Pacific Northwest Laboratory, TIAX LLC. (2008). *H2A hydrogen delivery infrastructure analysis models and conventional pathway options analysis results*. (Interim Report No. DE-FG36–05GO15032). Office of Energy Efficiency & Renewable Energy.
[21] Hourly Ontario energy price: independent electricity system operator. (2015). Retrieved 2 June, 2015 from http://www.ieso.ca/Pages/Power-Data/price.aspx
[22] Henry hub natural gas spot price: US energy information administration. (2015). Retrieved 2 June, 2015 from http://www.eia.gov/dnav/ng/hist/rngwhhdm.htm
[23] Market based regulation signals: ancillary services (PJM). (2014). Retrieved 1 December, 2014 from http://www.pjm.com/markets-and-operations/ancillary-services/mkt-based-regulation.aspx
[24] Regulation market capability clearing price: PJM. (2015). Retrieved 10 June, 2015 from http://www.pjm.com/markets-and-operations/billing-settlements-and-credit/preliminary-billing-reports/pjm-reg-data.aspx
[25] Transportation rates (M-12 rate schedule): Union gas. (2014).Union Gas.
[26] Plant cost index – Chemical Engineering. (2015). Retrieved 13 June, 2015 from http://www.chemengonline.com/pci-home
[27] *Rate Order 2014 Uniform Electricity Transmission Rates*. (2014). Ontario Energy Board.
[28] Demand side response in the ancillary service markets – Pennsylvania New Jersey Maryland interconnection. (2014). Retrieved (01/05/2015) from http://so-ups.ru/fileadmin/files/company/markets/dr/dr_pjm/dr_pjm_economy_rsu.pdf
[29] Ainscough, C., Peterson, D., and Miller, E. (2014). *Hydrogen production cost from PEM electrolysis*. DOE Hydrogen and Fuel Cells Program Record. Retrieved (15/05/2015) from: https://www.hydrogen.energy.gov/pdfs/14004_h2_production_cost_pem_electrolysis.pdf

Chapter 6

Methane in MOFs: where, why, and how

Trevor A. Makal[1]

6.1 Introduction

As atmospheric carbon dioxide (CO_2) levels continue to rise, concern over global climate change remains at the forefront of many discussions. As such, a large number of research topics involving conservation and remediation of the environment due to the impacts of atmospheric CO_2 have ballooned in the past decades, including capture of CO_2, conversion of CO_2 to useful products, and utilization of cleaner fuels, such as natural gas (NG) and hydrogen (H_2) [1–16].

In 2011, it was reported that 27 per cent of all greenhouse gas emissions were generated from transportation sources, which have been the fastest-growing source of US greenhouse gas emissions [17]. A paradigm shift involving a move away from petroleum to less carbon intensive fuel sources for vehicular use stands to substantially reduce the annual amount of anthropogenic CO_2 emissions. The two primary alternative fuels that have gained traction for on-board vehicular fuel systems are H_2 and NG. While H_2 is an ideal fuel, since it produces only water as a product of combustion, its application as an on-board vehicular fuel remains hindered by the lack of efficient storage methods and energy intensive production and isolation [18–20].

Methane, the primary component of NG (composing > 95 per cent), has a heat of combustion similar to that of gasoline (CH_4, 55.7 MJ/kg; gasoline, 46.4 MJ/kg) while boasting the smallest amount of CO_2 per unit of energy and being naturally abundant. Similarly to H_2, the discovery of efficient storage methods has prevented the widespread production and adoption of NG-fueled vehicles. Three primary methods of storing NG have been extensively investigated: compressed NG (CNG), liquefied NG (LNG), and adsorbed NG (ANG). In CNG processes, NG is pressurized to 200–300 bar at room temperature and achieves a volumetric energy density (VED) of 9.2 MJ/L (only 27 per cent of the VED of gasoline, 34.2 MJ/L). This significantly lower VED relates to reduced driving distances between fueling. Furthermore, at such high pressures heavy, specialized gas cylinders and multi-stage

[1]Department of Natural Science, The University of Virginia's College at Wise, Wise, VA 24293, USA

compressors are required, thereby decreasing useable space within the vehicle and increasing costs. LNG has significantly greater VED than CNG (22.2 MJ/L, 64 per cent of the VED of gasoline), but requires a temperature of −161.5 °C. Such low temperature necessitates the use of expensive cryogenic vessels and suffers from boil-off (which relates to both decreased driving distances and pronounced environmental concerns, as methane is a far more impactful greenhouse gas than CO_2). ANG, making use of a porous sorbent to enhance density of NG in a fuel tank, has been suggested as a "best-of-both-worlds" middle ground between CNG and LNG [21–23]. By incorporation of high surface area materials into a fuel tank, methane may favorably interact with the internal surfaces of these materials (adsorb) so as to realize high NG storage at pressures significantly lower than those required for CNG technologies (30–50 bar, typically). At these lower pressures, it is possible to avoid the use of heavy, thick-walled cylinders in favor of formable gas tanks similar to those currently used for liquid fuel storage. Current ANG technologies can achieve VEDs of approximately 6.9 MJ/L.

In 2012, the Advanced Research Projects Agency-Energy (ARPA-E) of the US Department of Energy issued a set of goals for NG storage (VED > 12.5 MJ/L) [24]. Of the four projects funded through the Methane Opportunities for Vehicular Energy program for development of new storage media, three involved the use of adsorbent materials. Of these sorbent-based projects, two focused on the application of advanced porous materials (metal-organic frameworks [MOFs] and porous organic polymers). MOFs comprise metal ions or atoms linked by multitopic organic linkers (independently referred to as secondary building units [SBUs]) to form extended two- or three-dimensional porous materials, oftentimes with high crystallinity [25]. The enormous variety in options involving organic and inorganic building unit selection gives rise to a plethora of possible structures with varying functionalities. The ability to tailor pore shapes and sizes, along with the presence of various functional groups has given rise to an incredible number of reports involving application of MOFs in fields such as gas storage, separations, catalysis, sensing, and drug delivery [26–35]. The past several years have produced a surge in research investigating methane storage in MOFs, and have led to several quality review papers providing excellent insight into the status of MOFs for ANG technologies [12, 13, 36, 37]. As such, it is the purpose of this mini review to provide an overview of the adsorption of methane in MOFs with particular focus on identification and efficacy of adsorption sites.

6.2 Identification of adsorption sites

In order to design functional materials for ANG technologies it is beneficial to identify the locations of methane molecules within porous materials so as to determine effective functionalities and pore geometries and sizes. A difficulty in determining these binding sites lies in the fact that methane adsorption in MOFs is dominated by weak van der Waals (vdW) interactions with isosteric heats of adsorption (ΔH_{ads}) in the range of 10–20 kJ/mol. To this end, various techniques have been used to predict and identify preferred adsorption, or binding, sites upon methane loading. Some of the techniques that have been employed to accomplish this task include neutron diffraction, Grand Canonical Monte Carlo (GCMC) simulations, density functional theorem (DFT) calculations, spectroscopic methods, and X-ray diffraction (XRD). Paired with gas sorption analysis, and resultant heats of adsorption data, these techniques have provided significant insight into the methane adsorption process.

Since MOFs tend to be crystalline materials (though not a requirement based upon recent IUPAC recommendations) [25], they may be easily modeled and studied by diffraction techniques and computational simulations to investigate structure as well as guest binding. Due to the complexity of MOF crystal structures and resultant disorder that arises from guest adsorption, diffraction experiments often require complementary computational analyses as a means to decipher diffraction patterns. Powder neutron diffraction has proved to be extremely useful in the identification of host–guest interactions in adsorbent materials. The Zhou group at National Institute of Standards and Technology (NIST) reported the analysis of methane adsorption sites in a series of MOFs, the M-MOF-74, or M_2dobdc series (M = Mg, Mn, Co, Ni, Zn; H_2dobdc =2,5-dihydroxyterephthalic acid), by powder neutron diffraction measurements paired with DFT calculations [38]. A key feature of the M-MOF-74 series is that they contain one-dimensional linear chains of metal ions connected by carboxylates and hydroxyl groups from the ligands (Figure 6.1(a) and (c)). Each metal comprising this linear chain is octahedrally coordinated with one site occupied by a solvent species. This solvent species may be removed through a variety of methods to generate a coordinatively unsaturated metal center (UMC). In this study, they identified that, similarly to dihydrogen uptake in MOFs, the UMCs are the primary binding sites for methane adsorption (Figure 6.1(b)).

While the high density of UMCs in the M-MOF-74 series effectively explains the high methane capacities in those materials, it was necessary to identify

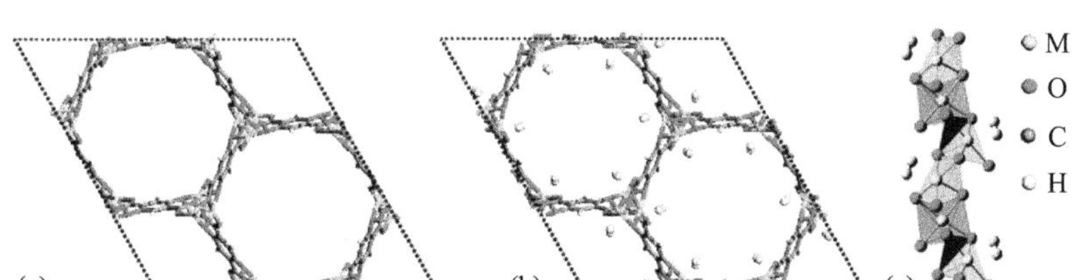

Figure 6.1 (a) [001] view of the unit cell of M_2dobdc. (b) The unit cell with H_2 molecules on the open metal sites. (c) A close view of the metal oxide polyhedrons, which form one-dimensional chains along the crystal c-axis. The metal ions are coordinately unsaturated, resulting in a distorted pyramid configuration. The H_2 molecule strongly adsorbed on the metal site partially serves the role of the missing ligand in a stable octahedral coordination. Reprinted with permission from Reference 39. Copyright 2009 American Chemical Society

alternative adsorption sites in other high uptake materials with much lower density of UMCs. The Zhou group selected a series of MOFs based upon dicopper paddlewheel and multicarboxylate organic linkers to identify other strong adsorption sites [40]. By selecting MOFs based upon the same metal SBU, evaluation of topological variations between the selected MOFs could be conducted by minimizing chemical variation between the studied materials. The MOFs chosen for this additional study included the well-known HKUST-1 (Cu_3btc_2, H_3btc = 1,3,5-benzenetricarboxylic acid), PCN-11 (Cu_2sbtc, H_4sbtc = *trans*-stilbene-3,3′,5,5′-tetracarboxylic acid), and PCN-14 (Cu_2adip, H_4adip = 5,5′-(9,10-anthracenediyl)-diisophthalic acid). A combination of powder neutron diffraction, GCMC simulations, and DFT calculations confirmed the role of UMCs as primary methane adsorption sites with highest heats of adsorption. The low density of UMCs in the materials, however, leads to only a minor contribution of these binding sites to the overall storage capacities. Upon increased loading of methane into the materials, it was identified that accessible small cages and channels led to enhanced vdW interactions and significant methane uptake. Importantly, it has become clear that adsorption sites that are very different in terms of energetics at low temperatures may be far less defined from one another at elevated temperatures. This has been proven especially true in the evaluations of PCN-14 and HKUST-1 toward methane sorption, where HKUST-1 has been identified to exhibit nonspecific binding of methane (i.e. methane guest species are found uniformly distributed throughout the system) and PCN-14 has exhibited similar behavior in which no energy barrier was found between the UMCs and secondary binding sites [40].

Single-crystal XRD analysis has been performed on methane-loaded MOF $Zn_2(bdc)_2(dabco)$ (H_2bdc = terephthalic acid, dabco = 1,4-diazabicylco[2.2.2]octane) [41]. The $Zn_2(dbc)_2(dabco)$ MOF is a quintessential example of a pillared-layer MOF in which the $Zn_2(COO)_4$ paddlewheel forms a two-dimensional sheet in the *ab* plane with the linear terephthalate linker, while dabco units serve as pillars along the *c*-axis to connect these two-dimensional sheets. Due to the occupancy of dabco on the terminal position of the paddlewheel, no UMCs may be generated in these types of structures. Kim *et al.* identified that at low temperatures (90 K) the MOF has three distinct methane binding sites present in the system (Figure 6.2) [41]. Two primary binding sites were identified, one near the $Zn_2(COO)_4$ paddlewheel SBU (A sites, Figure 6.2(c)) and the other at the center of the small windows along the *a* and *b* axes (B sites, Figure 6.2(d)). Sites A and B were found to be 0.5 occupancy in the crystal structure, indicative of a lack of preference between the two sites. The secondary adsorption sites (denoted C sites) are located at the centers of the pores and are 69 per cent occupied. Methane molecules in site A interact not only with the $Zn_2(COO)_4$ SBU but also exhibit CH-π interactions with the adjacent phenyl rings of the bdc ligands. Site B methane molecules interact with the sides of the phenyl rings of bdc ligands through vdW interactions. Methane molecules found in the secondary sites, C, located in the center of the pores are in close contact with sites A and B, and are stabilized by vdW interactions with those sites.

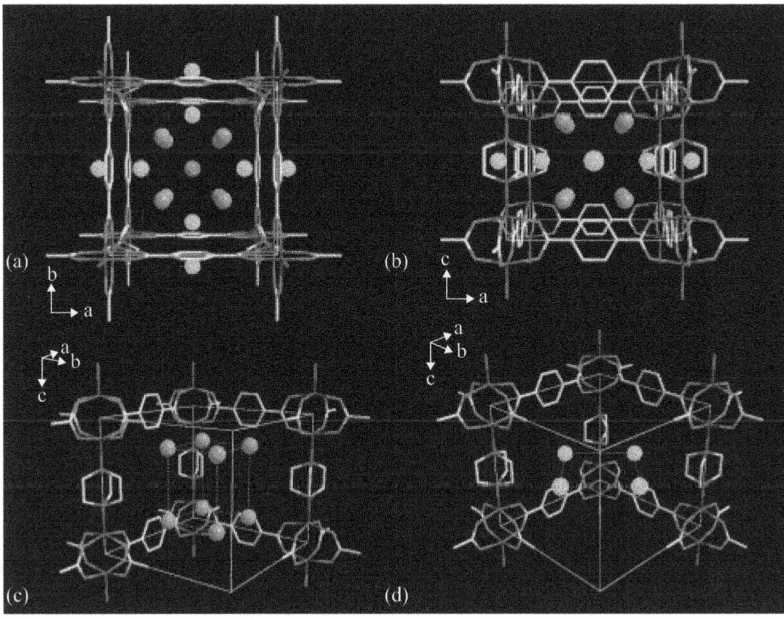

Figure 6.2 *(a) X-ray crystal structure of **1**·6.69CH$_4$ with three methane adsorption sites (A, B, and C) along the c-axis, (b) side view of the framework along the b-axis, (c) adsorption sites A, and (d) adsorption sites B (site A = light grey, site B = medium grey, site C = dark grey in center of cell). Reprinted with permission from Reference 41. Copyright 2009 John Wiley and Sons*

Spectroscopic techniques, including infrared (IR) spectroscopy and Raman spectroscopy have been extensively used in the probing of weak vdW interactions in gas adsorption applications, particularly H_2 and CO_2 adsorption processes [42–45]. IR spectroscopy evaluates absorption of IR radiation due to changes in vibrational levels in the electronic ground state of a molecule, whereas Raman spectroscopy probes changes in polarization of a molecule through electronic absorption of ultraviolet, visible, or near IR radiation. Though a particular free-guest species may be either IR or Raman inactive, the adsorption onto a surface may induce a decrease in symmetry of the adsorbate, leading to the presence of active modes. This has been shown to a great extent in the study of H_2 adsorption in MOFs, in particular. For identification of methane adsorption sites by spectroscopic methods, Raman spectroscopy has been most prevalent (though still quite underused).

In 2007, the Matzger group made use of Raman spectroscopy to probe the adsorption of methane in a series of MOFs with varying pore sizes but same underlying topology, dubbed the isoreticular MOF (IRMOF) series [46]. In particular, the authors were interested in investigating adsorption sites under conditions relevant to storage systems. While previous studies making use of inelastic neutron scattering (INS) and XRD had identified UMCs as primary binding sites, those studies had all been conducted at low temperatures, far below those applicable for on-board vehicular fuel system technologies. The chosen materials for analysis (IRMOF-1, -6, -8, -11, and -18) are all of the form Zn_4OL_3, where L is a linear dicarboxylate organic linker. The Zn_4O forms an octahedral SBU upon coordination with six carboxylates resulting in the IRMOF series adopting simple cubic structures. The authors identified that there are two distinct locations in which methane may adsorb on the surfaces of the MOF, namely the metal SBU and the organic linker. Previously conducted INS studies on H_2 sorption in IRMOFs indicated that the gas preferentially adsorbs near the zinc oxide SBU at low partial pressures with sites near the organic linker being occupied only after metal sites are saturated [47, 48]. Interestingly, in the Raman spectroscopic studies of methane adsorption in IRMOFs, significantly different vibrational shifts ($\Delta\nu = \nu_{free} - \nu_{adsorbed}$) were found for the adsorption of methane (ranging from 7.6 to 11.0 cm^{-1}), whereas frequency shifts upon N_2 adsorption in the same study varied by only 1.5 cm^{-1} [46]. This result indicates that methane adsorption in IRMOFs must be dependent upon organic linker, as no variation in guest binding would be expected if the zinc oxide SBU was the preferential adsorption site (the metal SBU being identical in all MOFs in the study). Currently, no further studies have been reported investigating why this may be the case, specifically why methane shows no preference toward the metal SBU while H_2, N_2, and Ar all preferentially occupy the site near the metal SBU over the linker site in this series of MOFs. It is possible that the larger size of methane makes it more difficult to interact at the zinc oxide SBU. Since each zinc is tetrahedrally coordinated, and no UMCs are available to directly bind guest species, the coordination environment around the metal would have to distort to a great extent to accommodate any significant binding, otherwise interactions near the metal center would be weak and similar in strength to that of sites near the organic linkers. It is important to consider, then, the heats of adsorption of targeted gases to materials in the IRMOF series. Hydrogen heats of adsorption in IRMOF series materials remain fairly constant upon increased loading, indicating that the energetics of adsorption at zero-loading (initial binding sites being occupied) are not significantly greater than upon secondary binding sites being occupied, comparable to adsorption on porous carbons [16]. Methane adsorption in IRMOF-1 behaves in a similar fashion, with a constant, rather low, ΔH_{ads} value of 12.3 kJ/mol, indicating that no strong binding sites are present in the material [36]. By comparison, M-MOF-74 (M_2dobdc) materials show a marked decrease in ΔH_{ads} upon increased CH_4 loading owing to the high density of UMCs in the materials [36]. The significant variation in binding in IRMOF-8 and -11 requires alternative explanation to describe the reason for enhanced binding affinity. IRMOF-11 is known to form a two-fold interpenetrated network that has been proposed to serve to enhance interaction with guest species due to increased overlap of potential energy surfaces between host and guest. Furthermore, it has been shown that IRMOF-8 can form an interpenetrated network, which would assist in the explanation of the significant differences observed in the Raman spectroscopy studies of methane binding in IRMOFs [49]. Most importantly, it is clear that the lack of UMCs in the IRMOF series results in primary binding sites being determined by proximity of organic linkers to one another to generate surfaces with increased potential overlap.

Powder XRD (PXRD) is used constantly in MOF chemistry to identify phase purity, but has been difficult to utilize for identification of host–guest interactions. Considering the difficulties in solving crystal structures from PXRD data, it is not difficult to understand the limited application of PXRD in identification of sorption sites. Recently, a new technique has been reported for quick generation of structure envelopes from difference envelope density (or differential electron density) analysis of MOFs to aid in structure solution and analysis from PXRD data [50]. Using an initial model of a MOF, determined from single-crystal XRD analysis, an ideal structure envelope may be generated. For any MOF materials with identical structures, the structure envelopes will be the same. With this in mind, the difference between observed (ρ_{obs}) and calculated ($k\rho_{calc}$ with scale factor, k) envelope densities, ρ_{Δ}, may be determined and used to identify the presence of guest species (equation inset in Figure 6.3). In Figure 6.3, a pictorial representation of this

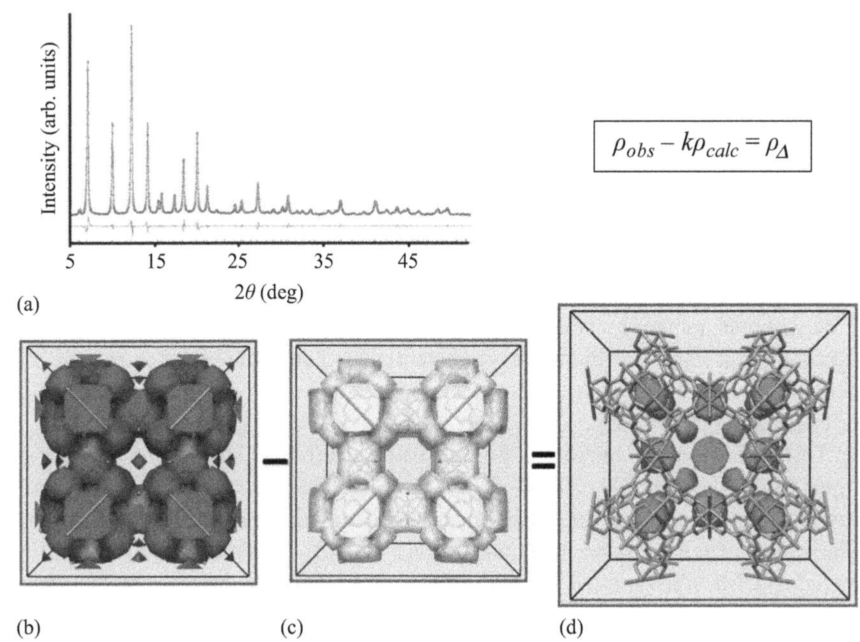

Figure 6.3 (a) Final Pawley whole pattern decomposition plot of HKUST-1, (b) its observed structure envelope based on ρ_{obs}, (c) its calculated structure envelope based on ρ_{calc}, and (d) its difference envelope density ρ_{Δ} overlapped with its structural model. Reprinted with permission from Reference 50. Copyright 2014 American Chemical Society

process is shown for an as-synthesized sample of HKUST-1 [50]. The residual density observed in the pores is representative of both coordinated and free residual solvent molecules present in the system. Significantly, the authors were able to conduct such evaluations using home-source diffractometers (Cu K_α radiation). Yakovenko *et al.* made use of this technique in MOF analysis to identify activation of, and CO_2 binding in, several MOFs [51].

This technique has been further applied to study real-time gas adsorption in MOFs [52]. In the first example of experimental crystallographic measurement of gas migration, Chen *et al.* applied time-resolved diffraction structure envelope (TRDSE) method to identify the evolution of gas binding sites within four MOFs with respect to temperature and exposure time to methane and CO_2. From their study of HKUST-1, the authors observed that methane is adsorbed in the small octahedral cages at 295 K, at the window openings at 200 K, and around the large cage corners at 150 K. The authors report that these results are consistent with the ensemble-averaged maps predicted by the Monte Carlo simulation in Materials Studio, as well as expectations based upon previous identification of binding site priorities in HKUST-1 (Figure 6.4). Interestingly, methane was found to initially bind to UMCs and then migrate to secondary adsorption sites. This phenomenon is further described in section 6.3.

6.3 Evaluation of adsorption sites

Adsorption sites in MOFs are typically described in order of occupancy and saturation. For a typical MOF with UMCs, the UMCs are described as the primary adsorption sites, followed by small cages and window sites classified as secondary adsorption sites. Depending on functionality and, likely more importantly, geometry, ligands may play a role as secondary or tertiary adsorption sites. The filling of large pores generally occurs through the process of absorption, rather than adsorption, but may contribute fairly substantially to absolute methane storage. An analysis of each of these sites is provided later in this section with an emphasis on the interplay between each.

UMCs have been identified as the primary binding sites for a number of targeted guest species [38, 39, 53–55]. This arises from the unscreened Coulomb attraction between metal ions and guest methane molecules resulting in slight polarization of the methane molecule (Figure 6.5) [38]. This enhanced interaction has led to UMCs being cited for high ΔH_{ads} at zero-loading in most MOFs which exhibit high initial heats of adsorption.

The significant role of UMCs in methane adsorption was exemplified in the study of M-MOF-74 (M_2dobdc, M = Mg, Mn, Co, Ni, Zn) materials in which >90 per cent of the experimental methane adsorption may be accounted for based upon loading onto UMCs in the MOFs [38]. It was identified that saturation of UMCs in these MOFs could contribute to volumetric methane storage capacities of 160–174 cm^3(STP)/cm^3. The Ni derivative, Ni-MOF-74, exhibited the highest experimental isosteric heat of adsorption and excess methane adsorption of the

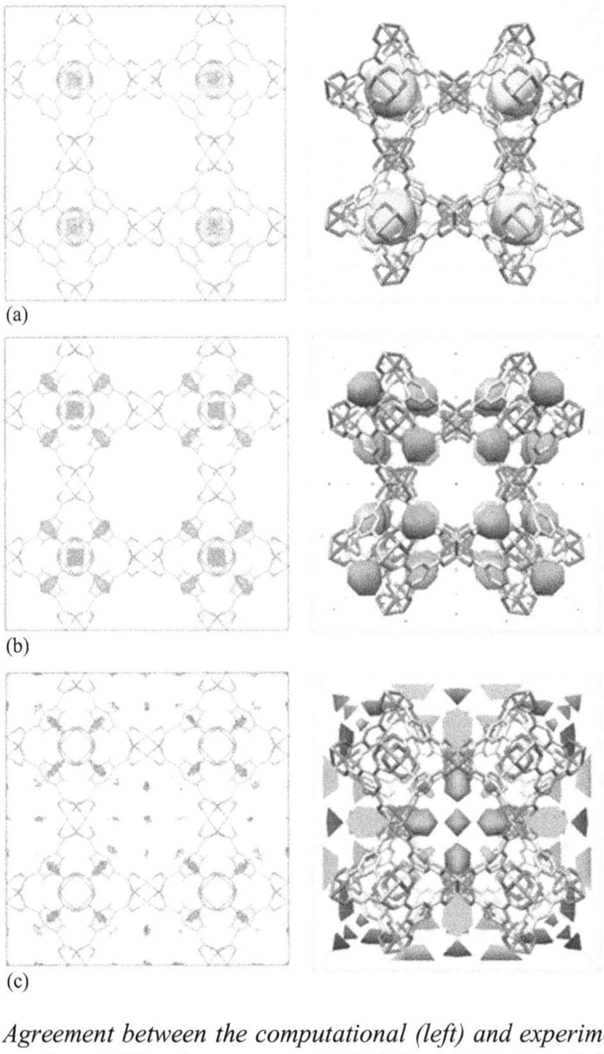

Figure 6.4 Agreement between the computational (left) and experimental (right) results of CH_4 adsorption at (a) 295 K, (b) 200 K, and (c) 150 K in HKUST-1. Reprinted with permission from Reference 52. Copyright 2015 American Chemical Society

series, 20.2 kJ/mol and 190 cm^3(STP)/cm^3 at 298 K and 35 bar, respectively. The Mg and Mn derivatives both exhibited experimental methane capacities lower than those predicted from full saturation of UMCs. The most likely explanation for such variation from predicted methane capacity is incomplete activation of the metal SBUs resulting in generation of fewer UMCs than expected. Incomplete activation of MOFs has been identified as a significant source of diminished performance of MOFs and has led to a large variety of activation conditions being explored.

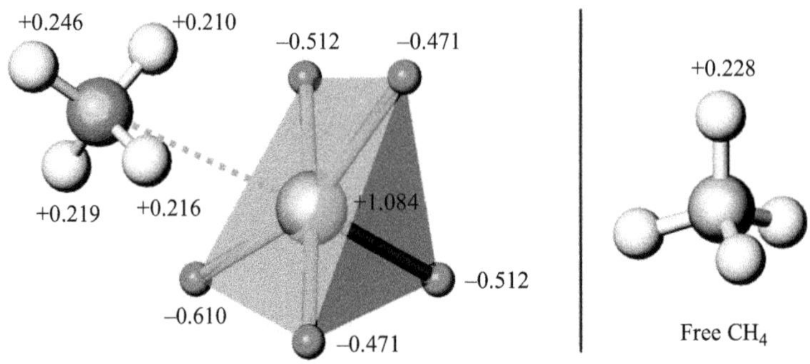

Figure 6.5 Projected Löwdin charges of the methane molecule adsorbed on the open Mg site in Mg-MOF-74 (Mg_2dobdc). The adsorbed CH_4 is slightly distorted with an uneven charge distribution, which enhances the electrostatic interaction with the open metal charges. For comparison, the Löwdin charge of the H atom in a free CH_4 gas molecule is also shown. Reprinted with permission from Reference 38. Copyright 2009 American Chemical Society

One technique that has gained significant fame in the past several years is the application of supercritical CO_2 to effectively wash the pores of the MOFs and remove all uncoordinated and loosely coordinated guest species [56]. Other methods of activation of MOFs to generate UMCs include exchanging high boiling point solvents often used in the synthesis process for lower boiling point and/or volatile solvents and/or evacuating the system under vacuum with slight heating. While UMCs can be used to enhance overall methane storage capacity, they primarily adsorb (and desorb) methane at low pressures which results in unusable fuel for on-board vehicular fuel systems and, therefore, low working capacities (defined as the difference between the amount of methane stored at the upper working pressure, typically 35 or 65 bar, and the lower working pressure, typically 5 bar). For example, Ni-MOF-74 has an experimental absolute methane storage capacity of 250 cm^3 (STP)/cm^3 at 298 K and 65 bar. However, even considering the application of slight heat to enhance delivery, the working capacity (65 bar at 298 K → 5 bar at 323 K) is only 165 cm^3(STP)/cm^3 [57]. Designing a tank system that would release all methane bound to metal sites would require significant heating or application of vacuum, feats that would likely result in complicated engineering of fuel tanks as well as parasitic energy loss. This trade-off is a well-known problem that exists in the majority of storage and separation applications of MOFs, highlighting the benefit of MOFs with fewer UMCs which engage in weaker interactions with methane molecules.

In another study, it has been proposed that the UMCs in PCN-14 play a directing role at relatively higher temperatures [58]. In particular, that computational analysis has shown that the strong and weak binding sites in PCN-14 identified from low temperature studies were found to have no energy barrier between them at elevated temperatures. Instead, a cooperative binding effect was

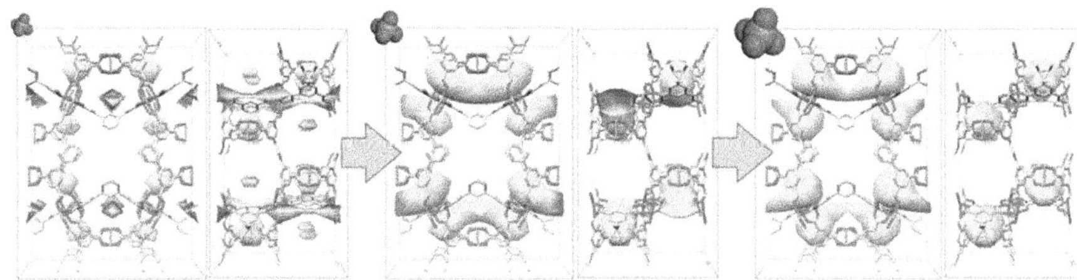

Figure 6.6 DED snapshots illustrate the kinetic transition in the CH_4 adsorption process in PCN-306. The perspectives are viewed along the c-axis (left) and a-axis (right). The CH_4 flow was introduced into the crystals, followed by sequential X-ray measurements after (a) 150 s, (b) 330 s, and (c) 510 s, under controlled conditions of 295 K and 1 bar. Clouds represent the most significant adsorption regions with contouring at 1.5σ. Reprinted with permission from Reference 52. Copyright 2015 American Chemical Society

observed in which the UMCs initially interact with incoming methane molecules and then direct them to nearby "weak" binding sites. Until recently, this computationally derived phenomenon had not been observed in experiment. However, with the advent of TRDSE, Chen *et al.* [52] were able to observe the evolution of gas binding sites within MOFs as methane was loaded into the systems. Figure 6.6 illustrates the observed kinetic transition of methane molecules upon adsorption into PCN-306 (Cu_2TPTA, H_4TPTA = [1,1′:3′,1″-terphenyl]-3,3″,5,5″-tetracarboxylic acid). Upon initial loading of methane in all three of the MOFs included in their study, the UMCs are first occupied; however, upon additional loading of methane, a transition is observed in which methane move from the UMCs to small window sites (Figure 6.6(c)). The authors suggest that the methane molecules undergo a "kinetic transition" from the UMCs to small window sites to reach a minimum Gibbs free energy for the adsorption process by increasing entropy. The localized-to-mobile mechanism supports the observations of molecular transport from one site to another and provides great insight into the role of UMCs in methane adsorption processes [52]. Interestingly, while adsorption occurs in a relatively indiscriminate fashion, desorption of methane occurs such that the gas remains on the UMCs, indicating the effects of the strong electrostatic forces on delivery of the gas (Figure 6.7).

While HKUST-1 is composed of dicopper-paddlewheel SBUs and possesses a fairly high density of UMCs, GCMC studies had shown that at 3.2 bar and 298 K methane molecules are uniformly dispersed throughout the unit cell, rather than occupying UMCs [59]. From TRDSE studies, it was also found that no preference for UMCs was present at 1 bar and 295 K, with methane occupying the small octahedral cages instead [52]. Furthermore, it was determined that upon lowering the temperature of the system to 200 K led to complete rearrangement of methane in which molecules were found to adsorb solely at the UMCs with limited occupancy of the large pores, as well. Upon lowering the temperature to 150 K, the

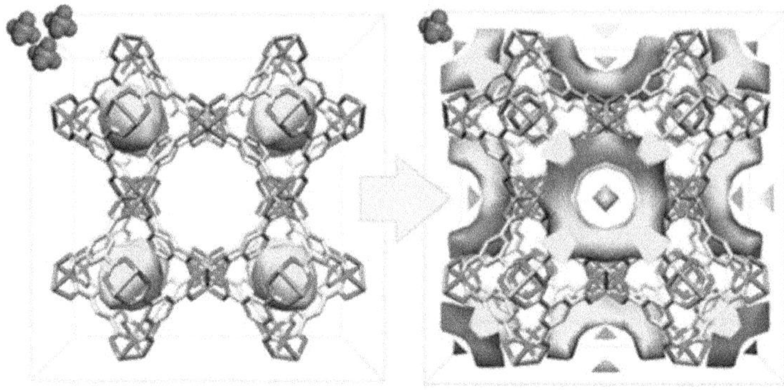

Figure 6.7 DED snapshots illustrate the evolution of CH_4 desorption in HKUST-1 at 295 K. The experiment procedure was controlled as follows: (a) CH_4 loading for 1500 s and (b) followed by He flushing for 300 s. Adapted from Reference 52. Copyright 2015 American Chemical Society

small octahedral cages remain unoccupied, but electron density attributed to methane is found in the larger pores. Chen *et al.* suggest that the decreased kinetic energy of methane at lower temperatures prevents overcoming the energy barrier created by H_2 atoms located near the window openings to smaller cages, thereby leading to a selectivity in which methane is only found in the small cages when sufficient energy is applied to "push through" the small windows. This requirement of methane to "force its way" through the energy barrier to access the small pores may be partially responsible for the recently discovered high storage and delivery capacities of HKUST-1, 227 cm^3(STP)/cm^3 total uptake capacity and 190 cm^3(STP)/cm^3 deliverable [57]. It is currently unclear how common such phenomena, in which UMCs play directing role rather than primary binding site, occur in MOFs but it is clear that UMCs are responsible for incomplete release of methane from MOFs upon desorption.

Another high methane capacity MOF, UTSA-20 ($Cu_3(bhb)_2$, H_6bhb = 3,3′,3″,5,5′,5″-benzene-1,3,5-triyl-hexabenzoic acid), with dicopper paddlewheel SBUs and easily generated UMCs has been thoroughly studied and reported [60]. UTSA-20 was found to adsorb methane up to 178 cm^3(STP)/cm^3 at room temperature and 35 bar with isosteric heat of adsorption at zero-loading comparable to that for other $Cu_2(COO)_4$ paddlewheel based MOFs with UMCs. The authors identified that saturation of UMCs could only contribute to approximately half of the total storage capacity and, therefore, expanded their study to include computational analysis of the structure to identify other binding sites. GCMC calculations identified that phenyl rings on the bhb^{6-} linkers were found to have short distances between one another, resulting in porous channels very near the size of one adsorbed methane molecule (Figure 6.8). The close proximity of ligand surfaces leads to increased potential overlap with adsorbed methane molecules, creating favorable methane adsorption sites.

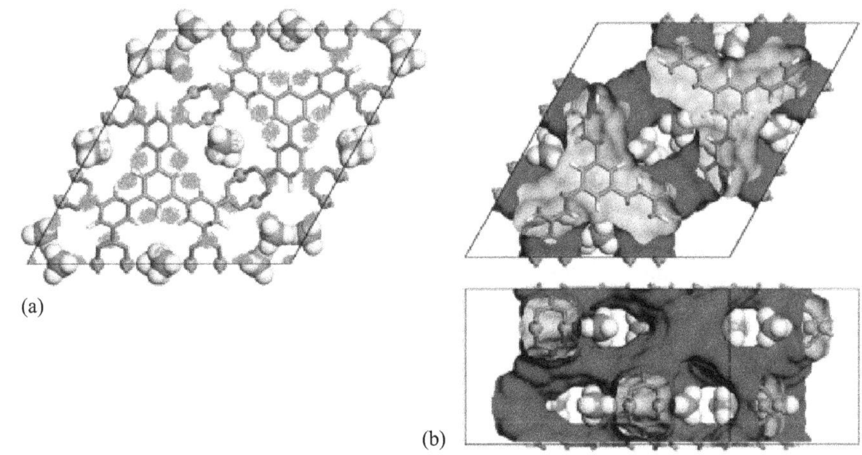

Figure 6.8 (a) Probability distribution of the CH_4 center of mass in UTSA-20 ([001] view), obtained from GCMC simulation at 298 K and 10 bar. The dotted regions represent the places where methane molecules are heavily populated in the MOF structure. (b) The pore surface of the interconnected channel pores in UTSA-20 with adsorbed methane at the linker channel site. The channel width along the c-axis matches well with the size of the adsorbed methane molecules, leading to enhanced vdW interaction (methane molecules are shown in space-filling representation for clarity). Reprinted from Reference 60. Copyright 2011 John Wiley and Sons

A simplified method for computationally screening MOFs for methane adsorption capabilities has led to the generation of a Hypothetical Metal-Organic Framework (HMOF) database designed and maintained by the Snurr lab [61], and has provided an opportunity to analyze design factors which are predicted to enhance methane capacity. Throughout their studies they determined that optimally performing MOFs tended to have channels defined by large, planar aromatic surfaces, and pore sizes around ~4 or ~7.75 Å, approximately 1–2 times the kinetic diameter of methane (3.8 Å). Further evaluation of their HMOF database concluded that lipophilic groups (methyl, ethyl, propyl) were common in HMOFs with high adsorption capacities [61]. It has been proposed that these types of functionalities may result in enhanced interactions with methane through nonpolar interactions. This type of interaction, however, may be too weak to significantly contribute to methane adsorption and may, rather, lead to decreased methane capacities due to pendant functionalities extending into the pores of the MOFs and limiting surface area and pore volume. Currently, no experimental evidence has been reported indicating enhanced interactions between MOF and methane due to nonpolar interactions. The majority of reports in which alkyl substituents have been shown to enhance gas sorption may be explained by simple geometric and size restraints. For instance, PCN-306 and its methyl substituted derivative, PCN-307, have been evaluated for methane adsorption sites [52]. While PCN-307 does exhibit increased methane adsorption located near the methyl groups, this site also serves as an exit from the pore in both PCN-306 and -307. Since the methyl groups effectively block a portion of the pore window, the smaller cross section of the exit may be accredited with preventing gas leakage (Figure 6.9). Additional work concerning identification of the potential role of lipophilic groups as methane adsorption sites is currently underway.

Large pores, while not serving as adsorption sites, nonetheless contribute to absolute methane capacity. Large pores in MOFs can serve to reduce density, contributing to higher gravimetric capacity and, conversely, lower volumetric capacity, as well as provide large volumes in which methane may absorb and enhance diffusion kinetics. In these regions of space, no significant host–guest interactions are present and, therefore, methane–methane interactions provide the most significant contributions to intermolecular interactions. As such, these volumes have been identified as very weak sorption sites and are generally regarded as detrimental to high methane capacities. It has generally been accepted that these sites are only filled after all strong adsorption sites have been saturated. Recently, however, it has been identified that gaseous guest species begin to occupy these regions of space long before all strong sites have been saturated [51]. Instead of a one-by-one filling order, multiple sites (strong and weak) were identified to be occupied at the same time, prior to all strong sites being occupied. While large pores may begin to be occupied prior to saturation of strong sites, they are still observed to be vacated first and maintain a lower density of methane than on pore surfaces.

6.4 Outlook and future directions

A very recently reported comprehensive study involving the Materials Genome Initiative has investigated in silico methane uptake in 650 000 materials based upon results of molecular simulation [62]. Through this study they were able to evaluate prominent features of materials with high calculated methane deliverable capacity and the practicality of the current ARPA-E deliverable capacity target. Perhaps the most important find in their report is related to this latter subject. In their analysis, it was determined that the current ARPA-E target for methane deliverable capacity of 315 v(STP)/v appears likely unachievable. This expansive study found that the highest predicted deliverable capacity in the set of 650 000 materials is 196 v(STP)/v, significantly lower than the ARPA-E target. Furthermore, the current highest deliverable capacity from experimental studies are 185 v(STP)/v (MOF-5 and HKUST-1), 194 v(STP)/v (USTA-76), and 208 v(STP)/v (MOF-519). The authors also note that it is striking that, even given possible errors in force fields, experimental sample quality, and experimental adsorption measurements, all experimental and computational work seem to converge on a maximum deliverable capacity of 200 v(STP)/v [62]. As such, it appears that current technologies may be at the peak of ANG potential.

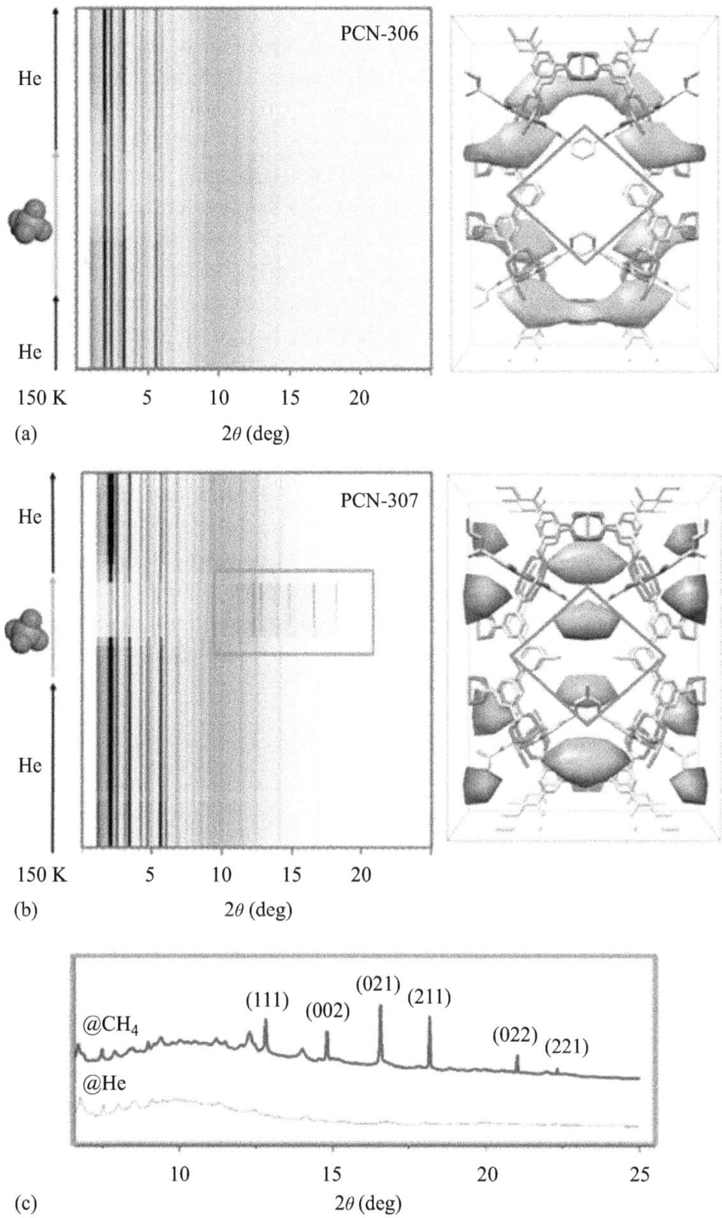

Figure 6.9 *(Left of a, b) The top view of the in situ gas-loading-dependent PXRD patterns for PCN-306/PCN-307 at 150 K. (Right of a, b) The corresponding DED maps for CH_4 adsorption. (c) The confined methane phase grown in PCN-307 at 150 K ($\lambda = 0.72910$ Å). Reprinted with permission from Reference 52. Copyright 2015 American Chemical Society*

Certainly, attempts to design new materials for methane storage will not cease. New materials may still be discovered with lower cost of production, higher sustainability, and lower environmental impact (including new methods for synthesizing MOFs using limited or no solvent). In the future design of materials for methane fuel delivery systems, it is important to bear in mind those bases determined from previous works. Based upon currently available information regarding efficacy of different adsorption sites in the development of sorbent materials for vehicular methane fuel storage, it appears clear that the design of new materials should include porous cages and channels tailored so as to maximize potential overlap between pore walls and guest methane species, and to limit the density of UMCs. This can be accomplished by designing materials with channel sizes near integer multiples of the kinetic diameter of methane (3.8 Å). Analysis from the Materials Genome Initiative seems to indicate that a maximum pore size of approximately 11 Å is ideal for maximizing deliverable capacity in porous materials.

Further analysis of methane sorption phenomena in MOFs will shed more light on principles of adsorption in highly porous materials and provide significant insight into design of new materials, not just for methane delivery applications but also other potential applications. Additional evaluation of diffusion processes involved in the flow of guests species through the porous channels and cages, along with kinetic studies involving individual binding sites, will aid in understanding the processes by which molecules move and react in small spaces. Continued development of new techniques for analysis of such phenomena is expected to expand greatly on our knowledge of such systems.

Acknowledgment

The author would like to extend gracious thanks to Dr Andrey A. Yakovenko for discussion particularly pertaining to application of PXRD in analysis of guest molecules in MOFs.

References

[1] Candelaria S.L., Shao Y., Zhou W., *et al.* 'Nanostructured carbon for energy storage and conversion'. *Nano Energy.* 2012;**1**(2):195–220.

[2] Liu J., Thallapally P.K., McGrail B.P., Brown D.R., Liu J. 'Progress in adsorption-based CO_2 capture by metal–organic frameworks'. *Chemical Society Reviews.* 2012;**41**(6):2308–2322.

[3] Li J-R., Sculley J., Zhou H-C. 'Metal–organic frameworks for separations'. *Chemical Reviews.* 2012;**112**(2):869–932.

[4] Dawson R., Cooper A.I., Adams D.J. 'Chemical functionalization strategies for carbon dioxide capture in microporous organic polymers'. *Polymer International.* 2013;**62**(3):345–352.

[5] Cheung O., Hedin N. 'Zeolites and related sorbents with narrow pores for CO_2 separation from flue gas'. *RSC Advances.* 2014;**4**(28):14480–14494.

[6] Pera-Titus M. 'Porous inorganic membranes for CO_2 capture: present and prospects'. *Chemical Reviews.* 2014;**114**(2):1413–1492.

[7] Gargiulo N., Pepe F., Caputo D. 'CO_2 adsorption by functionalized nanoporous materials: a review'. *Journal of Nanoscience and Nanotechnology.* 2014:**14**(2):1811–1822.

[8] Yang L., Wang H. 'Recent advances in carbon dioxide capture, fixation, and activation by using N-heterocyclic carbenes'. *ChemSusChem.* 2014; **7**(4):962–998.

[9] Wang J., Huang L., Yang R., *et al.* 'Recent advances in solid sorbents for CO_2 capture and new development trends'. *Energy & Environmental Science.* 2014;**7**(11):3478–3518.

[10] Leung D.Y.C., Caramanna G., Maroto-Valer M.M. 'An overview of current status of carbon dioxide capture and storage technologies'. *Renewable & Sustainable Energy Reviews.* 2014;**39**:426–443.

[11] Taheri Najafabadi A. 'Emerging applications of graphene and its derivatives in carbon capture and conversion: current status and future prospects'. *Renewable & Sustainable Energy Reviews.* 2015;**41**:1515–1545.

[12] Makal T.A., Li J-R., Lu W., Zhou H-C. 'Methane storage in advanced porous materials'. *Chemical Society Reviews.* 2012;**41**(23):7761–7779.

[13] He Y., Zhou W., Qian G., Chen B. 'Methane storage in metal–organic frameworks'. *Chemical Society Reviews.* 2014;**43**(16):5657–5678.

[14] Sreenivasulu B., Gayatri D.V., Sreedhar I., Raghavan K.V. 'A journey into the process and engineering aspects of carbon capture technologies'. *Renewable & Sustainable Energy Reviews.* 2015;**41**:1324–1350.

[15] Wang C-C., Zhang Y-Q., Li J., Wang P. 'Photocatalytic CO_2 reduction in metal–organic frameworks: a mini review'. *Journal of Molecular Structure.* 2015;**1083**(5):127–136.

[16] Collins D.J., Zhou H-C. 'Hydrogen storage in metal–organic frameworks'. *Journal of Materials Chemistry.* 2007;**17**(30):3154–3160.

[17] U.S. Environmental Protection Agency, *Inventory of U.S. greenhouse gas emissions and sinks: 1990–2011.* 2013. EPA 430-R-13–001.

[18] Schlapbach L., Züttel A. 'Hydrogen-storage materials for mobile applications'. *Nature.* 2001;**414**:353–358.

[19] Aricò A.S., Bruce P., Scrosati B., Tarascon J-M., van Schalkwijk W. 'Nanostructured materials for advanced energy conversion and storage devices'. *Nature Materials.* 2005;**4**:366–377.

[20] Murray L.J., Dincă M., Long J.R. 'Hydrogen storage in metal–organic frameworks'. *Chemical Society Reviews.* 2009;**38**(5):1294–1314.

[21] Düren T., Sarkisov L., Yaghi O.M., Snurr R.Q. 'Design of new materials for methane storage'. *Langmuir.* 2004;**20**(7):2683–2689.

[22] Menon V., Komarneni S. 'Porous adsorbents for vehicular natural gas storage: a review'. *Journal of Porous Materials.* 1998;**5**(1):43–58.

[23] Matranga K.R., Myers A.L., Glandt E.D. 'Storage of natural gas by adsorption on activated carbon'. *Chemical Engineering Science.* 1992; **47**(7):1569–1579.

[24] Advanced Research Projects Agency – Energy United States Department of Energy, *Methane opportunities for vehicular energy*. 2012. Funding Opportunity Number DE-FOA-0000672.

[25] Batten S.R., Champness N.R., Chen X-M., *et al.* 'Terminology of metal–organic frameworks and coordination polymers'. *Pure and Applied Chemistry.* 2013;**85**(8):1715–1724.

[26] Kuppler R.J., Timmons D.J., Fang Q-R., *et al.* 'Potential applications of metal–organic frameworks'. *Coordination Chemistry Reviews.* 2009; **253**(23–24):3042–3066.

[27] Kreno L.E., Leong K., Farha O.K., Allendorf M., Van Duyne R.P., Hupp J.T. 'Metal–organic framework materials as chemical sensors'. *Chemical Reviews.* 2012;**112**(2):1105–1125.

[28] Lee J., Farha O.K., Roberts J., Scheidt K.A., Nguyen S.T., Hupp J.T. 'Metal–organic framework materials as catalysts'. *Chemical Society Reviews.* 2009;**38**(5):1450–1459.

[29] Kitagawa S., Kitaura R., Noro S. 'Functional porous coordination polymers'. *Angewandte Chemie International Edition.* 2004;**43**(18):2334–2375.

[30] Czaja A.U., Trukhan N., Müller U. 'Industrial applications of metal–organic frameworks'. *Chemical Society Reviews.* 2009;**38**(5):1284–1293.

[31] James S.L. 'Metal–organic frameworks'. *Chemical Society Reviews.* 2003;**32**(5):276–288.

[32] Férey G. 'Hybrid porous solids: past, present, future'.*Chemical Society Reviews.* 2008;**37**(1):191–214.

[33] Chen B., Xiang S., Qian G. 'Metal–organic frameworks with functional pores for recognition of small molecules'. *Accounts of Chemical Research.* 2010;**43**(8):1115–1124

[34] Corma A., García H., Llabrés i Xamena F.X. 'Engineering metal organic frameworks for heterogeneous catalysis'. *Chemical Reviews.* 2010;**110**(8): 4606–4655.

[35] Li J-R., Kuppler R.J., Zhou H-C. 'Selective gas adsorption and separation in metal–organic frameworks'. *Chemical Society Reviews.* 2009;**38**(5): 1477–1504.

[36] Mason. J.A., Veenstra M., Long J.R. 'Evaluating metal–organic frameworks for natural gas storage'. *Chemical Science.* 2014;**5**(1):32–51.

[37] Adisa O.O., Cox B.J., Hill J.M. 'Methane storage in molecular nano-structures'. *Nanoscale.* 2012;**4**(11):3295–3307.

[38] Wu H., Zhou W., Yildirim T. 'High-capacity methane storage in metal–organic frameworks M_2(dhtp): the important role of open metal sites'. *Journal of the American Chemical Society.* 2009;**131**(13): 4995–5000.

[39] Zhou W., Wu H., Yildirim T. 'Enhanced H_2 adsorption in isostructural metal–organic frameworks with open metal sites: strong dependence of the binding strength on metal ions'. *Journal of the American Chemical Society.* 2008;**130**(46):15268–15369.

[40] Wu H., Simmons J.M., Liu Y., *et al.* 'Metal–organic frameworks with exceptionally high methane uptake: where and how is methane stored?'. *Chemistry – A European Journal.* 2010;**16**(17):5205–5314.

[41] Kim H., Samsonenko D.G., Das S., *et al.* 'Methane sorption and structural characterisation of the sorption sites in $Zn_2(bdc)_2(dabco)$ by single crystal X-ray crystallography'. *Chemistry – An Asian Journal.* 2009;**4**(6):886–891.

[42] Brown C.M., Ramirez-Cuesta A.J., Her J-H., Wheatley P.S., Morris R.E. 'Structure and spectroscopy of hydrogen adsorbed in a nickel metal–organic framework'. *Chemical Physics.* 2013;**427**(12):3–8.

[43] Bordiga S., Regli L., Bonino F., *et al.* 'Adsorption properties of HKUST-1 toward hydrogen and other small molecules monitored by IR'. *Physical Chemistry Chemical Physics.* 2007;**9**(21):2676–2685.

[44] Greenaway A., Gonzalez-Santiago B., Donaldson P.M., *et al.* 'In situ synchrotron IR microspectroscopy of CO_2 adsorption on single crystals of the functionalized MOF $Sc_2(BDC\text{-}NH_2)_3$'. *Angewandte Chemie International Edition.* 2014;**53**(49):13483–13487.

[45] Chen Y., Wang H., Li J., Lockard J.V. 'In situ spectroscopy studies of CO_2 adsorption in a dually functionalized microporous metal–organic framework'. *Journal of Materials Chemistry A.* 2015;**3**(9):4945–53.

[46] Siberio-Perez D.Y., Wong-Foy A.G., Yaghi O.M., Matzger A.J. 'Raman spectroscopic investigation of CH_4 and N_2 adsorption in metal–organic frameworks'. *Chemistry of Materials.* 2007;**19**(15):3681–3695.

[47] Rowsell J.L.C., Eckert J., Yaghi O.M. 'Characterisation of H_2 binding sites in prototypical metal–organic frameworks by inelastic neutron scattering'. *Journal of the American Chemical Society.* 2005;**127**(42): 14904–14910.

[48] Rosi N.L., Eckert J., Eddaoudi M., *et al.* 'Hydrogen storage in microporous metal–organic frameworks'. *Science.* 2003;**300**(5622):1127–1129.

[49] Feldblyum J.I., Dutta D., Wong-Foy A.G., *et al.* 'Interpenetration, porosity, and high-pressure gas adsorption in Zn_4O(2,6-naphthalene dicarboxylate)$_3$'. *Langmuir.* 2013;**29**(25):8146–8153.

[50] Yakovenko A.A., Wei Z., Wriedt M., Li J-R., Halder G.J., Zhou H-C. 'Study of guest molecules in metal−organic frameworks by powder X-ray diffraction: analysis of difference envelope density'. *Crystal Growth & Design.* 2014;**14**(11):5397–5407.

[51] Yakovenko A.A., Reibenspies J.H., Bhuvanesh N., Zhou H-C. 'Generation and applications of structure envelopes for porous metal–organic frameworks'. *Journal of Applied Crystallography.* 2013;**46**(2):346–353.

[52] Chen Y-P., Liu Y., Liu D., Bosch M., Zhou H-C. 'Direct measurement of adsorbed gas redistribution in metal–organic frameworks'. *Journal of the American Chemical Society.* 2015;**137**(8):2919–2930.

[53] Xiang S., Zhou W., Zhang Z., Green M.A., Liu Y., Chen B. 'Open metal sites within isostructural metal–organic frameworks for differential recognition of acetylene and extraordinarily high acetylene storage capacity at

room temperature'. *Angewandte Chemie International Edition*. 2010;**49**(27): 4615–4618.

[54] Bloch E.D., Queen W.L., Krishna R., Zadrozny J.M., Brown C.M., Long J.R. 'Hydrocarbon separations in a metal–organic framework with open Iron(II) coordination sites'. *Science*. 2012;**235**(6076):1606–1610.

[55] Dietzel P.D.C., Johnsen R.E., Fjellvåg H., *et al*. 'Adsorption properties and structure of CO_2 adsorbed on open coordination sites of metal–organic framework Ni_2(dhtp) from gas adsorption, IR spectroscopy and X-ray diffraction'. *Chemical Communications*. 2008;(48):5125–5127.

[56] Farha O.K., Hupp J.T. 'Rational design, synthesis, purification, and activation of metal–organic framework materials'. *Accounts of Chemical Research*. 2010;**43**(8):1166–1175.

[57] Peng Y., Krungleviciute V., Eryazici I., Hupp J.T., Farha O.K., Yildirim T. 'Methane storage in metal–organic frameworks: current records, surprise findings, and challenges'. *Journal of the American Chemical Society*. 2013;**135**(32):11887–11894.

[58] Lucena S.M.P., Mileo P.G.M., Silvino P.F.G., Cavalcante Jr. C.L. 'Unusual adsorption site behavior in PCN-14 metal–organic framework predicted from Monte Carlo simulation'. *Journal of the American Chemical Society*. 2011;**133**(48):19282–19285.

[59] Karra J.R., Walton K.S. 'Effect of open metal sites on adsorption of polar and nonpolar molecules in metal–organic framework Cu-BTC'. *Langmuir*. 2008;**24**(16):8620–8626.

[60] Guo Z., Wu H., Srinivas G., *et al*. 'A metal–organic framework with optimized open metal sites and pore spaces for high methane storage at room temperature'. *Angewandte Chemie International Edition*. 2011; **50**(14):3178–3181.

[61] Wilmer C.E., Leaf M., Lee C.Y., *et al*. 'Large-scale screening of hypothetical metal–organic frameworks'. *Nature Chemistry*. 2012;**4**(2):83–89.

[62] Simon C.M., Kim J., Gomez-Gualdron D.A., *et al*. 'The materials genome in action: identifying the performance limits for methane storage'. *Energy Environmental Science*. 2015;**8**(4):1190–1199.

Chapter 7

Thermal management as a key in storing adsorbed natural gas

Mojtaba Ahmadi-Baloutaki[1]*, Rupp Carriveau*[1] *and David S-K. Ting*[1]

7.1 Introduction

Recent improvements in natural gas extraction coupled with environmental and economic benefits of natural gas as an inexpensive and clean vehicular fuel have driven a renewed interest in natural gas vehicles [1–3]. Along with this interest is the recognition that the world's natural gas resource base is both abundant and under-utilized [4]. Therefore, many industries have focused on using natural gas as their primary fuel in the recent decades. This will increase the natural gas demand significantly in the future. For instance, Solomon Natural Gas Services reported that the gas demand for Alberta oil sands will have a great increase in ten years whereas it will be just over 2 per cent of the total North American demand in 2025 [5]. On the other hand, the high costs associated with natural gas transportation and reduced energy density at standard temperature and pressure conditions are considered as the major obstacles against its commercialization at a larger scale [3]. There are a few ways to store natural gas in automotive industry including liquefied natural gas (LNG), compressed natural gas (CNG) and adsorbed natural gas (ANG). Natural gas stored as an adsorbed phase in porous materials is referred to as ANG which is the topic of this chapter with a particular focus on vehicular applications. This storage method as a new technology has some advantages over CNG and LNG.

One of the primary advantages of ANG over CNG storage is its lower storage pressure (2–4 MPa) compared to the high pressure (20 MPa) required for CNG. This lower storage pressure reduces both the cost of compression stations and that of the storage vessel since lighter vessels can be utilized for adsorptive storages [6]. The ANG storage is also less of a safety hazard than CNG due to the lower operating pressure. Moreover, a lower operating pressure allows the utilization of more conformable storage tanks. The current CNG tanks are cylindrical with semi-spherical caps and their conformability factor, defined as the ratio of outer tank volume to the

[1]Turbulence and Energy Laboratory, Centre for Engineering Innovation, University of Windsor, Windsor, Ontario N9B 3P4, Canada

enclosing rectangular cuboid volume, is about 50–60 per cent [7]. This results in some unused space when the gas storage is installed within the trunk of a vehicle. On the other hand, non-cylindrical tanks having conformability factors as high as 90 per cent can be used to accommodate the relatively low pressure of ANG storage.

ANG is also a preferred method for storing natural gas over LNG. This is mainly because of the fact that the operating temperature range of ANG is the room temperature while LNG is usually stored at low temperatures of about 112 K (–161 °C) in cryogenic tanks [8]. The pressure storage of LNG is at atmospheric pressure. The low temperature associated with LNG tanks results in the requirement of costly liquefaction storage and periodic venting [4]. Moreover, refueling must be handled by trained personnel due to the danger of possible injury through contact with cryogenic liquids. This is, however, not an issue in ANG refueling process.

Apart from the aforementioned advantages of ANG, designing an ANG storage vessel with high storage capacity to provide an acceptable mileage range is crucial for further development of this technology in automotive industry. Currently, a basic ANG storage design, considered as a storage vessel filled with the adsorbent material, can store up to 190 volumes of gas per volume of storage space (v/v) at pressures of approximately 35 bar. This storage density, however, is only about 20 per cent of that of gasoline. On the basis of an identical fuel tank volume, switching to ANG indicates reduced driving range which is undesirable for long-range driving. On the other hand, ANG provides less pollution and extracting more mileage per dollar [4]. Although this may seem a trade-off situation, the necessity for further research on several aspects of ANG technology to increase the storage capacity is inevitable.

There is ongoing research on finding high performance adsorbents to increase the storage density of ANG tanks [9–13]. This part of ANG research focuses on highly micro-porous adsorbents with relatively high packing densities and large surface area to maximize the gas storage. This topic will not be discussed in this chapter anymore and interested reader is referred to additional references [9–13]. In addition to utilizing an effective adsorbent, the thermal and flow characteristics of the ANG storage system should be optimized with the goal of achieving higher storage. Most of the basic ANG systems designed have the storage and delivery capacities of about 85–90 per cent lower than those obtained on an ideally isothermal process. This capacity loss depends mainly on the shape of the adsorption isotherm curves and thermal effects during dynamic operation. The ANG thermal effects depend on the heat of adsorption, the heat transfer properties of the adsorbent bed and also on the heat exchanged through the walls of the vessel and with the surroundings. Consequently, it can be concluded that the performance and viability of an ANG system depends not only on the effectiveness of the adsorbent, but also on the heat and mass transfer properties of the adsorbent bed and the storage vessel [6].

Another vital requirement for commercializing ANG technology is the need for short refueling time. The refueling time of less than 5 min seems practical in order to make ANG a competitive solution to other vehicular fuels. However, achieving this short refueling time is usually accompanied by sacrificing a portion of the maximum achievable storage capacity due to the heat released during the adoption. In fact, adsorption is an exothermic process and the temperature of the adsorbent increases during adsorption. This temperature increase results in the reduction of gas adsorption which will be explained in details in section 7.2. Therefore, the minimization of thermal effects during adsorption is considered as a key in developing ANG vessels for commercial use. Effective thermal management not only increases the gas storage density, but it also results in shorter filling time. This chapter discusses the fundamental aspects of thermal management applicable to ANG storage vessels.

7.2 Adsorption thermodynamics

Designing an effective thermal management system for ANG tanks requires a sound understanding of thermodynamic process of adsorption. A brief description of adsorption thermodynamics is presented in this section. Further details can be found in more specialized texts [14–17]. Generally, adsorption is defined as the process in which the molecules of a gas are taken up by a solid surface. From a thermodynamics standpoint, during an adsorption process whereas a solid surface is exposed to a gas at pressure P in a closed space, the weight of the solid typically increases while the pressure of the gas decreases [14]. This pattern continues until a dynamic equilibrium is reached where the weight stops increasing and the pressure remains unchanged.

The basis for thermodynamic analysis is adsorption isotherms. The adsorption isotherm for a gas is the relation between the amount of gas adsorbed by the solid adsorbent with the adsorptive equilibrium pressure, while the temperature T and all other parameters are constant [15]. The isotherm functions are usually reported in terms of moles of gas adsorbed n per unit mass of the adsorbent bed. These isotherms are either measured experimentally or determined from molecular simulations.

The thermodynamic requirement for a spontaneous process indicates that the adsorption is an exothermic process. Considering the heat of adsorption being negative ($Q <$) by the formal sign convention for heat and work interactions, it can be concluded from the second law of thermodynamics described by (7.1), that the adsorption is accompanied by a decrease in entropy ($ds < 0$).

$$dS > \frac{\delta Q}{T} \tag{7.1}$$

The physical reason for this entropy reduction is that the degrees of freedom of the gas molecules in the adsorbed state are lower than those of in the gaseous state [14]. Using this thermodynamic analysis, the effect of the adsorption heat on the ANG storage can be discussed. The adsorption heat increases the temperature of adsorbent bed which results in a depression in adsorption process due to the enhanced mobility of the adsorptive molecules. This is why the adsorption heat is considered as a primary obstacle against gas storage. The effect of adsorption temperature on the amount of gas adsorbed during adsorption process is schematically depicted in Figure 7.1.

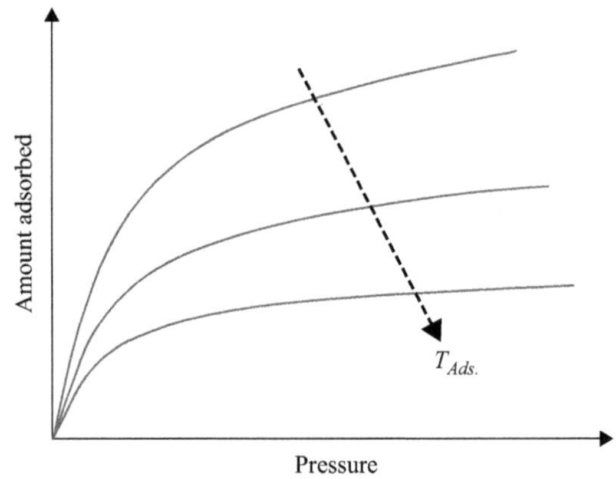

Figure 7.1 Effect of adsorption temperature on the amount of gas adsorbed

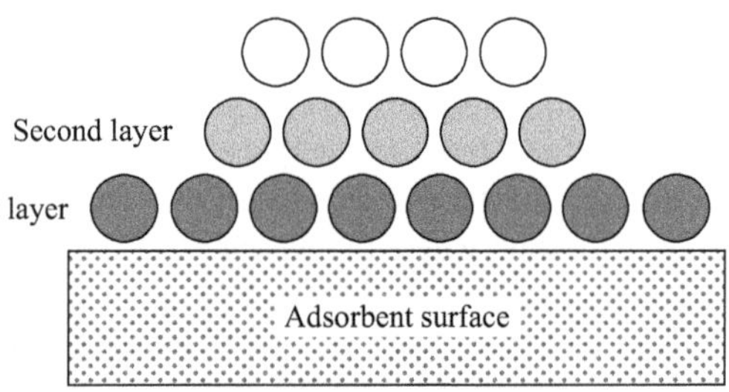

Figure 7.2 Schematic illustration of the multilayer pattern of adsorption

As can be observed, the amount of adsorbed gas is reduced when adsorption occurs at higher temperature.

Another impediment for adsorptive storage commercialization is the long filling time associated with ANG tanks as also mentioned in section 7.1. This aspect of adsorption process can also be explained from a thermodynamic standpoint. The surface atoms of a solid are responsible for the adsorption process. The adsorption, however, does not occur only on a single layer of the solid surface. In fact, adsorption is a multilayer process taking place on the surface of a solid as illustrated schematically in Figure 7.2. The fractional coverage of the adsorbate, Θ, at a particular level and at a given equilibrium pressure is defined as the ratio of the number of adsorbate particles that occupied available surface sites, N_a, to the total number of available adsorption sites, N_s, as follows [16]

$$\Theta = \frac{N_a}{N_s} \tag{7.2}$$

The adsorption at a particular level is considered complete when all the available adsorption sites at that level are occupied by the adsorbate particles, or in a mathematical form when $\Theta = 1$. Considering the multilayer pattern of adsorption, it can be stated that the suppressing effect of adsorption heat on the adsorption process is amplified significantly since the adsorption heat is influencing all the adsorption levels. Loss of storage capacity of ANG vessels was reported in the literature when the filling time was shortened. Braslaw *et al.* [18] reported 7 per cent lower methane storage capacity when the filling time decreased from overnight to 1 h. Faster fillings reported by Remick and Tiller [19, 20] (5 and 0.5 min fill, respectively) reduced the storage capacity by about 20 per cent and 25 per cent, respectively, compared to a slow overnight fill that allows the heat of adsorption to dissipate.

The insights gained from the thermodynamic analysis of the adsorption process should be kept in mind when a thermal management unit is to be designed. As a primary conclusion, it can be stated that the thermal management system should effectively influence the heat transfer mechanisms during the adsorption process between the adsorbate and the adsorbent particles on a multilayer pattern. This is the topic of the next section.

7.3 Thermal management approaches

As mentioned in the previous sections, thermal management is applied to ANG storage systems with the goal of increasing the storage density and shortening the refueling time. This is done by minimizing the thermal fluctuations of the storage bed during dynamic operation. The temperature increases as high as 80 °C have been measured during the charge cycle of an ANG storage model [21]. This calls for efficient methods of thermal management to overcome such high temperature fluctuations. Several approaches implemented by different researchers can be categorized into three groups as will be discussed later in the this section.

7.3.1 Increasing thermal capacity of the ANG storage system

The thermal capacity of the adsorbent will directly influence the heat transfer characteristics of the adsorbent system, the rate of charge or discharge and the temperature fluctuations in the adsorbent bed [4]. Using low thermal capacity adsorbents results in the significant temperature increase of the ANG storage system which accompanies by the lesser gas uptake. An adsorbent with high thermal capacity is an ideal choice for ANG storage since it helps mitigating the temperature fluctuation of the system. Therefore, one possible approach for the thermal management of ANG units is to artificially increase the thermal capacity of the ANG storage unit. One method proposed is using a phase-change material that can dissipate the adsorption heat. Toledo *et al.* [22] showed that it is possible to mitigate temperature fluctuation and achieve higher adsorption by using phase-change spheres inside the adsorbent bed. They placed 360 phase-change material spheres having 40 mm diameter inside an ANG vessel filled with granulated activated carbon and studied the thermal controlling effect of the phase-change material both computationally and experimentally. They reported that increasing the number of the phase-change material spheres inside the ANG tank increased the gas adsorption while reduced the adsorbent peak temperature, an indication of suppressing

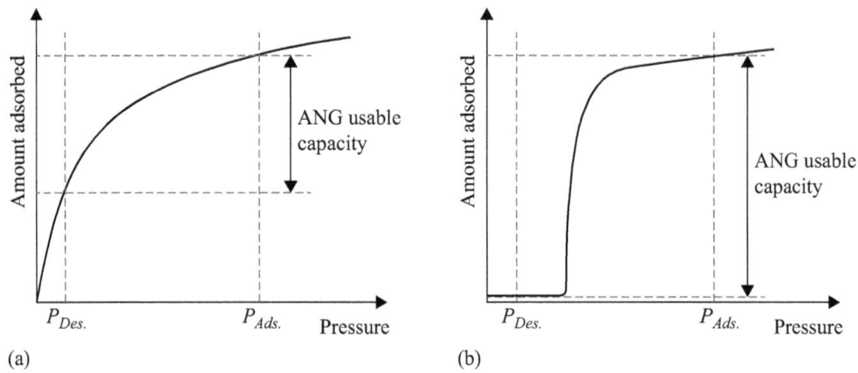

Figure 7.3 Schematic adsorption isotherms for (a) an adsorbent with S-shaped or stepped isotherm and (b) an adsorbent with a classical Langmuir-type isotherm

adsorption heat. These improvements, however, were made at a cost of losing a portion of storage volume. Therefore, researchers in this area have focused on finding new adsorbent materials with specific chemical composition that undergo phase-change during some operating pressures due to their unique structural bonds.

Following the same concept, Mason *et al.* [23] utilized a flexible compound of metal organic framework to maximize the deliverable capacity of methane. The particular chemical composition of the adsorbent they investigated was undergoing a structural phase transition after a certain threshold gas pressure is reached. The adsorption isotherm of this particular adsorbent has an S-shaped where the amount of gas adsorbed is small at low pressures but rise sharply just before the pressure reaches the desired storage pressure. This resulted in a significant increase in the usable capacity of ANG, which is defined as the difference between the amount of gas adsorbed at the storage pressure and the amount adsorbed at the lowest desorption pressure [23].

Figure 7.3 compares the adsorption isotherm for an adsorbent with S-shaped or stepped isotherm and an adsorbent with a classical Langmuir-type isotherm schematically. The adsorbed amount of methane reported by Mason *et al.* [23] is 197 v/v for adsorption at 65 bar, which are the highest values of usable methane capacity reported so far for any adsorbent under these conditions. In addition to the higher value of adsorbed gas, the approach implemented by Mason *et al.* [23] has the advantage of not occupying any extra space of the ANG storage vessel while former methods for increasing the adsorbents' thermal capacity is accompanied with reducing the volume available for storage.

7.3.2 Increasing thermal conductivity of the ANG storage system

Another popular way for controlling the thermal fluctuation of the adsorptive systems is to use the adsorbents having high thermal conductivity so that the heat of adsorption can be easily dissipated. In an effort to simulate the ideal characteristics of an adsorbent, Biloe *et al.* [24] showed that high conductivity is one of the required features for suppressing the adsorption heat using an analytical model for adsorption. Biloe *et al.* [24] also experimentally studied the effect of adsorbent's thermal conductivity on the ANG storage capacity using two different adsorbents with low and high thermal conductivities of 0.375 and 7.3 W/m K. The adsorption loss compared to the isothermal condition was less than 5 per cent with the highly conductive material while it was more than 20 per cent for the adsorbent with low thermal conductivity.

In another study on thermal conductivity of adsorbent materials, Zheng *et al.* [25] presented the results of their experiments on synthesizing a composite adsorbent based on four key parameters including heating temperature, heating time, mass ratio and consolidated pressure to maximize the adsorbent's thermal conductivity. Their results show that the maximum thermal conductivity of the composite occurs where the composite adsorbent was consisted of the equal amount of activated carbon and expanded graphite, heated at the temperature of 600 °C for the duration of 30 s, and consolidated under pressure 9 MPa. Zheng *et al.* [25] also showed that the thermal effect on the ANG storage vessel containing this synthesized adsorbent is much weaker compared to that of the tank with the base adsorbent. Goetz and Biloe [26] studied the heat transfer mechanisms between the adsorbent bed and the vessel walls. They demonstrated that the adsorption process of a highly conductive material occurs at almost isothermal conditions when combined with efficient heat transfer at the vessel wall.

7.3.3 Using a heat exchanger configuration

Although the two previous approaches of thermal management show promising results for suppressing the adsorption heat, another effective approach is utilizing a heat exchanger configuration inside the ANG storage unit. This approach is expected to be more efficient since it uses the forced advection as the dominating heat transfer mechanism between the gas and the adsorbent. It should be noted that the heat transfer is dominated by diffusion for the two previous approaches discussed earlier in this section. The major drawback of using a heat exchanger inside an ANG system is losing some storage space of the tank due to the presence of the heat exchanger components.

Various heat exchanger designs for minimizing the temperature fluctuations of ANG storage systems have been suggested [27–29]. Zheng *et al.* [27] conducted a series of experiments on an ANG storage system with thermal management composed of circulating water around the central region of the storage vessel and charging/discharging the gas through a perforated tube. They managed to control the temperature fluctuation of the vessel up to only 15 °C and 10 °C during the charge and discharge, respectively. Consequently, the amount of adsorbed/desorbed gas was increased by 14.2 per cent and 4.4 per cent, respectively [27]. Saez and Toledo [28] conducted a series of broad temperature measurements to study the thermal effects of adsorbent bed on the adsorption and desorption process of an ANG system. They measured the severest temperature fluctuation in the central region of the storage vessel (far from vessel walls) during both charge and discharge cycles. They also showed that the temperature fluctuations of the adsorbent bed strongly depend on the adsorption pressure as well as the adsorbent used. Saez and Toledo [28] suggested utilizing efficient heat transfer systems to eliminate the undesirable effects of adsorption heat. Rahman *et al.* [29] numerically modeled an ANG storage cylinder with fin and tube arrangement in the adsorbent bed taking the inertial and viscous effects of the heat and fluid flow through the porous adsorbent. Under the dynamic conditions, Rahman *et al.* [29] found that the charge period is shortened as the adsorbent temperature drops faster and the delivered amount of gas is maximized for the heat supply to the adsorbent bed during the discharge cycle.

Although heat exchanger configurations integrated within ANG storage vessels improve the thermal performance of ANG systems, they have some drawbacks on the design of ANG vessels by increasing the weight and the costs associated with the manufacturing and maintenance of these vessels. Therefore, designing a heat exchanging system that can effectively manage the thermal fluctuations of an ANG unit without further complicating the design of the ANG vessel is considered more efficient. A few heat exchanger designs that could benefit from different flow control techniques and efficiently influence the adsorption heat of ANG systems proposed in the literature [30–32]. Pupier *et al.* [30] proposed using a cycling scheme for storing ANG to mitigate the thermal effects of the adsorbent bed. In this scheme, the adsorbent was given some time to cool down after each cycle. They used a cylindrical vessel filled with the adsorbent material and equipped with a central gas diffuser for their experimental study. Pupier *et al.* [30] reported a decrease in the adsorbent mean temperature with increasing the cycle number. Moreover, the amount of time needed for the adsorbent to get back to the room temperature was shortened as the cycle number increased. In another flow control scheme on ANG systems, Santos *et al.* [31] proposed a recirculating design for filling the ANG storage vessels. In this design schematically depicted in Figure 7.4, the ANG tank has two opening ports, one as inlet and the other one as outlet. After starting the adsorption process, the non-adsorbed portion of the gas gets the chance to leave the storage vessel through the outlet port and enter a heat exchanger for cooling down before re entering the ANG vessel for adsorption while at a lower

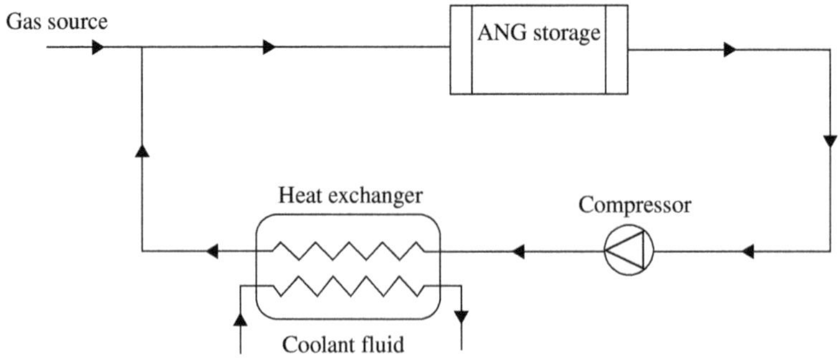

Figure 7.4 A recirculating scheme with thermal management for storing ANG

temperature. Santos *et al.* [31] numerically modeled this scheme and studied the thermal behavior of the ANG storage system during the charge cycle. Their model was included of an ANG vessel made up of several tubes, compacted with activated carbon with an external heat exchanger installed close to the gas source of the refueling station. Their design resulted in a filling time in the range of 50–200 s, depending on the mass flow rate, which is considered satisfactory for fast charge conditions thanks to the effective thermal management unit employed.

Following the recirculating filling scheme presented in Figure 7.4, Ahmadi-Baloutaki *et al.* [32] proposed a similar design with the potential advantages of a thermal management system while needless of a typical heat exchanger unit. Their proposed design shown in Figure 7.5 uses an ANG reservoir with much higher storage volume compared to the ANG vessel. This ANG reservoir can potentially function as a heat exchanging system by cooling down the high temperature gas exiting from the ANG vessel and entering to the reservoir. It seems that the designs discussed in Figures 7.4 and 7.5 increase the total cost of an ANG system by using additional components such as compressor, heat exchanger, ANG reservoir and extra tubing and fitting. Nonetheless, the potential saving from these thermal management schemes could be more beneficial by providing high performance ANG systems.

Another design for thermally managing the ANG systems by benefiting from flow control schemes was proposed by Ahmadi-Baloutaki *et al.* [32]. This design utilizes filter tube arrays submerged in the pack of metal organic framework powder. The filter tubes are tubes with porous walls that act as pairs of source and sink. The filter tubes connected to the injection port act as point sources injecting the gas to the metal organic framework during the charge cycle. The filter tubes connected to the collection port act as point sinks collecting the gas from the metal organic framework during the discharge cycle. These source and sink tubes can be connected to each other as proposed by others. In this particular design, however, the source and sink tubes are ended somewhere on the inside wall of the ANG vessel separately as depicted in Figure 7.6. This local blockage helps the gas

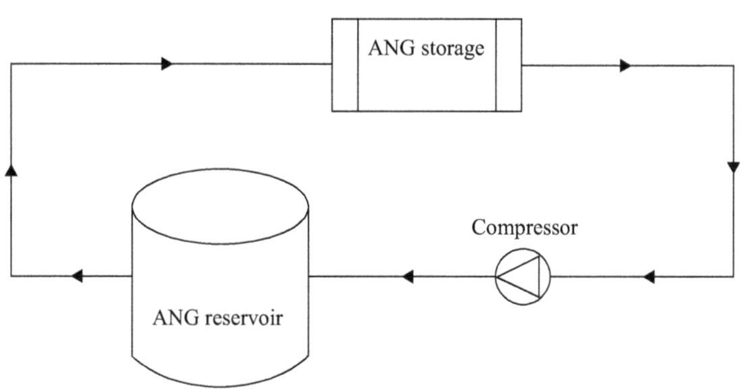

Figure 7.5 A recirculating scheme for storing ANG without a typical heat exchanger

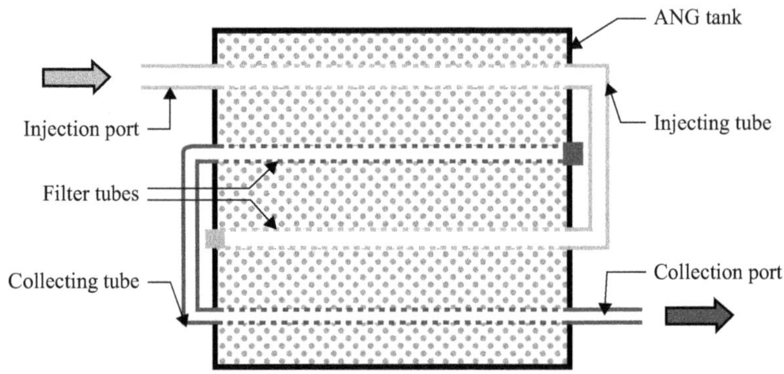

Figure 7.6 A schematic of the ANG storage design with filter tube arrays

flowing from and to the metal organic framework by the forced advection mechanism. This results in advective mass and heat transfer which is the key point in this design in that the adsorption heat can be dissipated with a higher rate. In addition to the increased rate of heat transfer from adsorbent bed by using the filter tubes, these tubes improve the structural strength of the storage vessel since they also function as the extra structural support. This allows using lighter vessels with smaller wall thickness. It should be noted that a major drawback of this tube configuration is the high pressure losses between injecting and collecting port which requires an external compressor to overcome this losses.

7.4 Conclusion

Due to the undesirable effects of adsorption heat on the adsorptive storage vessels, these storage systems are not efficient enough to utilize in the automotive industry currently. Despite the research on various aspects of the adsorptive storage, it appears that thermal, and accordingly flow, management is substantially effective in improving the performance of this type of storage for gaseous fuels especially natural gas. Fundamental aspects of an ideal thermal management unit were explained in this chapter from a thermodynamic standpoint. Several thermal management approaches proposed and tested in the literature were discussed. It was demonstrated that the applying thermal management to an adsorptive storage system results in two major benefits, namely higher gas storage and shorter filling time.

References

[1] Beckner M., Dailly A. 'A pilot study of activated carbon and metal–organic frameworks for methane storage'. *Applied Energy*. 2016; 162: 506–514.

[2] Rahman K.A., Chakraborty A., Saha B.B., Ng K.M. 'On thermodynamics of methane + carbonaceous materials adsorption'. *International Journal of Heat and Mass Transfer*. 2012; 55: 565–573.

[3] Yeh S., 'An empirical analysis on the adoption of alternative fuel vehicles: the case of natural as vehicles'. *Energy Policy.* 2007; 35: 5865–5875.
[4] Menon V.C., Komarneni S. 'Porous adsorbents for vehicular natural gas storage: a review'. *Journal of Porous Materials.* 1998; 5: 43–58.
[5] Mauger S. and Hensbergen G. '*Solomon forecasts gas demand for oil sands to 2025*'. Calgary, AB, Canada: Solomon Natural Gas Services, March 2016.
[6] Bastos-Neto M., Torres A.E.B., Azevedo D.C.S., Cavalcante Jr. C.L. 'A theoretical and experimental study of charge and discharge cycles in a storage vessel for adsorbed natural gas'. *Adsorption.* 2005; 11: 147–157.
[7] Beckner M., Dailly A. 'Adsorbed methane storage for vehicular applications'. *Applied Energy.* 2015; 149: 69–74.
[8] Judd R.W., Gladding D.T.M., Hodrien R.C., Bates D.R., Ingram J.P., Allen M. 'The use of adsorbed natural gas technology for large scale storage'. *ACS Division of Fuel Chemistry.* 1998; 43(3): 575–579.
[9] Garcia Blanco A.A., Vallone A.F., Korili S.A., Gil A., Sapag K. 'A comparative study of several microporous materials to store methane by adsorption'. *Microporous and Mesoporous Materials.* 2016; 224: 323–331.
[10] Makal T.A., Li J.-R., Lu W., Zhou H.-C. 'Methane storage in advanced porous materials'. *Chemical Society Reviews.* 2012; 41(23): 7761–7779.
[11] Simon C.M., Kim J., Gomez-Gualdron D.A., *et al.*, 'The materials genome in action: identifying the performance limits for methane storage'. *Energy & Environmental Science.* 2015; 8: 1190–1199.
[12] Bhatia S.K., Myers A.L. 'Optimum conditions for adsorptive storage'. *Langmuir.* 2006; 22(4): 1688–1700.
[13] Rodríguez-Reinoso F., Nakagawa Y., Silvestre-Albero J., Juárez-Galán J.M., Molina-Sabio M. 'Correlation of methane uptake with microporosity and surface area of chemically activated carbons'. *Microporous and Mesoporous Materials.* 2008; 115(3): 603–608.
[14] Bolis V. 'Fundamentals in adsorption at the solid–gas interface, concepts and thermodynamics'. in Auroux A. (Ed.). *Calorimetry and thermal methods in catalysis. Springer Series in Materials Science*, Vol. 154, Berlin Heidelberg: Springer-Verlag; 2013, pp. 3–50.
[15] Myers A.L., Monson P.A. 'Physical adsorption of gases: the case for absolute adsorption as the basis for thermodynamic analysis'. *Adsorption.* 2014; 20: 591–622.
[16] Myers A.L. 'Thermodynamics of adsorption'. in Letcher T.M. (Ed.). *Chemical thermodynamics for industry.* Royal Society of Chemistry; Cambridge, U.K., 2004. pp. 3–50.
[17] Saha P., Chowdhury S. 'Insight into adsorption thermodynamics'. in Tadashi M. (Ed.). *Thermodynamics.* InTech; Rijeka, Croatia, 2011, pp. 349–364.
[18] Braslaw J., Nasea Jr. J., Golovoy A. 'Low pressure methane storage system for vehicles-Preliminary concept evaluation'. *Proceedings of the 4th Miami Conference on Alternative Energy Sources*; Miami Beach, USA, 1981, pp. 261–270.

[19] Remick R.J. Tiller A.J. 'Heat generation in natural gas adsorption storage system'. *Proceedings of the Conference of Gaseous Fuels for Transportation I*; Vancouver, Canada, 1986, pp. 477–488.
[20] Remick R.J. Tiller A.J. 'Advanced methods for low-pressure storage of CNG'. *Proceedings of the Symposium on Nonpetroleum Vehicular Fuels V;* Arlington, USA, 1985, pp. 105–119.
[21] Chang K.J., Talu O. 'Behavior and performance of adsorption natural gas storage cylinders during discharge'. *Applied Thermal Engineering.* 1996; 16: 359–374.
[22] Toledo M., Rojas C., Montes E., Veloso J., Saez A. 'Use of phase change materials on an adsorbed carbon dioxide storage system'. *Applied Thermal Engineering.* 2013; 51: 512–519.
[23] Mason J.A., Oktawiec J., Taylor M.K., *et al.*, 'Methane storage in flexible metal organic frameworks with intrinsic thermal management'. *Nature.* 2015; 527(7578): 357–361.
[24] Biloe S., Goetz V., Guillot A. 'Optimal design of an activated carbon for an adsorbed natural gas storage system'. *Carbon.* 2002; 40: 1295–1308.
[25] Zheng Q.R., Zhu Z.W., Feng Y.L., Wang X.H. 'Development of composite adsorbents and storage vessels for domestically used adsorbed natural gas'. *Applied Thermal Engineering.* 2016; 98: 778–785.
[26] Goetz V., Biloe S. 'Efficient dynamic charge and discharge of an adsorbed natural gas storage system'. *Chemical Engineering Communications.* 2005; 192: 876–896.
[27] Zheng Q.R., Zhu Z.W., Wang X.H. 'Experimental studies of storage by adsorption of domestically used natural gas on activated carbon'. *Applied Thermal Engineering.* 2015; 77: 134–141.
[28] Saez A., Toledo M. 'Thermal effect of the adsorption heat on an adsorbed natural gas storage and transportation systems'. *Applied Thermal Engineering.* 2009; 29: 2617–2623.
[29] Rahman K.A., Loh W.S., Chakraborty A., Saha B.B., Chun W.G., Ng K.M. 'Thermal enhancement of charge and discharge cycles for adsorbed natural gas storage'. *Applied Thermal Engineering.* 2011; 31: 1630–1639.
[30] Pupier O., Goetz V., Fiscal R. 'Effect of cycling operations on an adsorbed natural gas storage'. *Chemical Engineering and Processing.* 2005; 44: 71–79.
[31] Santos J.C., Marcondes F., Gurgel J.M. 'Performance analysis of a new tank configuration applied to the natural gas storage systems by adsorption'. *Applied Thermal Engineering.* 2009; 29: 2365–2372.
[32] Ahmadi-Baloutaki M., Lee J.K., Carriveau R., Ting D.S-K. 'Experimental characterization of an adsorbed natural gas test cell'. *Proceeding of 2015 Natural Gas & Hydrogen Storage Symposium*; Windsor, Canada, 2015.

Chapter 8

Gas hydrate potential and development for methane storage

Ayush Goyal[1,2], Jacqueline Stagner[2] and David S-K. Ting[2]

8.1 Introduction

With the rise in population worldwide, the demand for energy is increasing exponentially. According to the World Energy outlook 2014, global energy demand will have a growth of 37% by 2040 [1]. Currently, these energy needs are mostly met by the extraction of fossil fuels. It is expected that unconventional sources (which include renewable resources) will be the major source of energy production but the energy generated from them might not be sufficient to meet all energy needs.

It has been found that the global demand for coal will grow by 15% by 2040 and the demand for natural gas is expected to grow by more than half its current amount. These figures clearly show the dependence on fossil fuels in coming decades (76% of energy comes from carbon sources gas/coal/oil) [1]. The burning and utilization of these fossil fuels lead to the emission of large amounts of carbon dioxide and other greenhouse gases to the atmosphere leading to increases in pollution and global warming. Not only this, it is expected that the total number of vehicles on the road will be doubled by the year 2040, which implies more emissions and, hence, more pollution [1]. Since natural gas is a cleaner burning fuel and creates less pollution compared to other fuels, it is expected that coal will be replaced by natural gas in the near future. Figure 8.1 shows how methane, which is highest constituent of natural gas, is less polluting than other fuels [2].

Gas hydrates are naturally occurring substances which contain a large quantity of methane per molecule of hydrate and are present in abundance in our earth. As shown in Figure 8.2, it has been estimated that world reserves of gas hydrates, both onshore and offshore, equal 10,000 gigatons (1 gigaton $= 10^9$ tons) [3]. These materials are considered an important source of natural gas for the future [4] due to their occurrence in the shallow geosphere [5]. Their potential as a future

[1]Turbulence and Energy Laboratory, Centre for Engineering Innovation, University of Windsor, Windsor, Ontario N9B 3P4, Canada
[2]Chemical Engineering Department, Indian Institute of Technology Roorkee, Roorkee Main Road, Uttarakhand, India

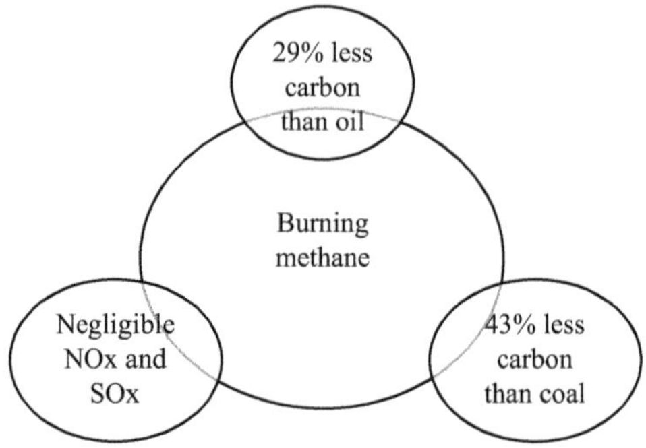

Figure 8.1 Emissions of NOx, SOx and carbon for methane compared to coal and oil [2]

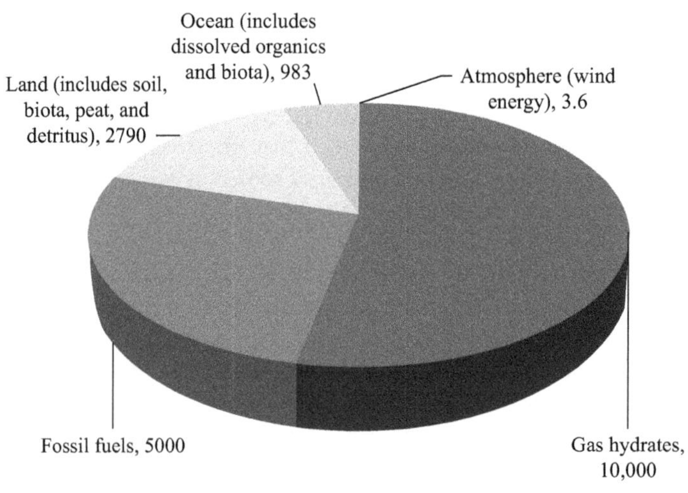

Figure 8.2 Available energy resources (in gigatons) (Modified from Reference 3)

energy source has not gone unnoticed. The governments of the United States, Japan, India, China, and other nations have all developed methane hydrate research and development programs [6]. As mentioned above, these hydrates provide methane which is a cleaner burning fuel compared to other fuels and, hence, cause less pollution. The economical extraction of methane from these natural gas hydrates is a key objective of present researchers. There are several ways in which the recovery of methane can be accomplished but they cannot be used on a large scale and are not cost effective. This will be one of the main points to ponder in the coming decades so as to ensure sufficient clean energy for the future [7, 8].

This chapter discusses the properties and provides a brief history of these hydrates. Also, it discusses available options for the recovery of methane and the current Indian scenario in this field. The future global perspective has also been mentioned.

8.2 Properties and composition of gas hydrates

Gas hydrates (or clathrate hydrates) are clathrate compounds that consist of crystalline molecules in which gas molecules physically occupy cavities within a crystal structure formed by water molecules. Water forms the host molecule lattice that traps guest molecules that may be a gas or a liquid [3, 4]. Gas hydrates are formed by a reaction of water with light natural gases such as carbon dioxide, methane, ethane, propane and butane [3]. Methane hydrates or natural gas hydrates are those clathrates in which this guest molecule is methane. These gas hydrates are solids similar to ice and are formed at low temperatures and high pressures (around 250 K and 25 MPa) [9]. The structure of a typical gas hydrate can be seen in Figure 8.3.

Methane hydrates have some unique properties which make them different from other resources. These include their stability at different temperatures and pressures. As shown in Figure 8.4, one volume of methane hydrate can contain up to 164 volumes of methane and 0.8 volume of water [10]. Its decomposition follows a rise in pressure. The composition of guest gases determines a hydrate's density and stability at various external conditions. Hydrates have high acoustic conductivity and low electrical conductivity. The composition of these hydrates mainly depends on how these hydrates were formed, that is, what were the external conditions (temperature and pressure), the composition of the guest gases and the location of formation. This composition might get altered by time since the conditions beneath the earth's surface might not remain consistent [11].

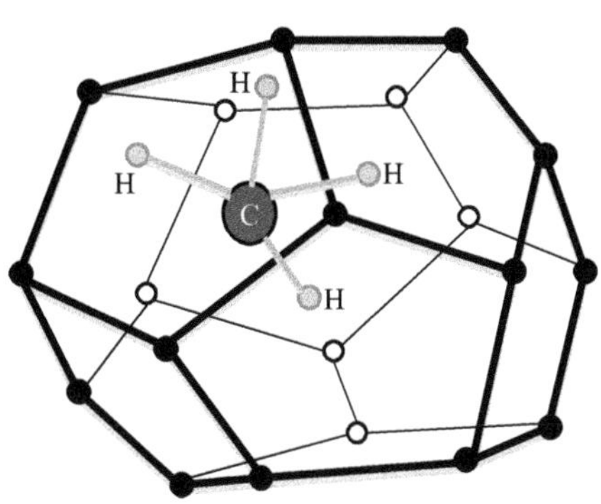

Figure 8.3 Typical structure of gas hydrate. Methane molecule is trapped inside the host water molecule (black dots)

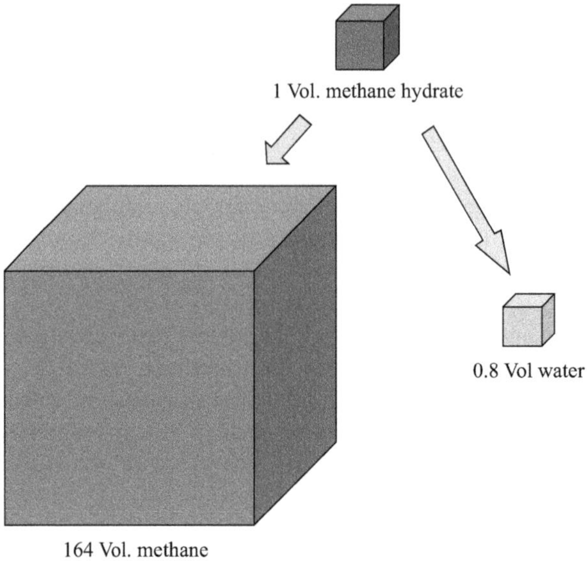

Figure 8.4 Methane hydrate decomposition on a volume basis

8.3 Brief history of gas hydrates

It has been more than a couple of centuries since gas hydrates became one of the main topics to ponder upon among researchers. The earliest hydrate formation is credited to Humphrey Davy and Michael Faraday in the 1800s [12]. Davy identified chlorine as an element and he, along with Faraday, continued experimenting with it. In 1823, Faraday reported a strange substance and called it a chlorine clathrate hydrate. It was 1888 when the first hydrocarbon hydrate was discovered by Paul Villard who synthesized a methane hydrate [12].

In the 1930s, cold temperature exposed pipeline clogging by ice-like material was reported. Scientists found that this material was ice wrapped around methane and developed ways to prevent its formation by glycols and methanols. Also, in the 1960s it was discovered for the first time that some gas hydrate deposits existed in the Messoyakha gas fields. Due to this discovery, many other places were explored for the existence of gas hydrates and many nations such as the United States, Japan, Canada, India and others started looking for these hydrate deposits [12].

Sloan [13] classifies three phases of gas hydrate history:

- The first period, from their discovery in 1810 until the present, includes gas hydrates as a scientific curiosity in which gas and water are transformed into a solid.
- The second period, continuing from 1934 until the present, concerns man-made gas hydrates as a hindrance to the natural gas industry.
- The third period, from the mid-1960s until the present, involves the discovery of the first hydrocarbon gas hydrates and studies on their utilization.

8.4 Location of gas hydrate deposits

A vast number of expeditions and explorations are being carried out for the assessment of various marine hydrate deposits all around the world. Their quantity, quality, properties, and stability zones are being researched and mapped. These estimates are made by direct analysis which involves drilling or indirect analysis which includes many seismic and geographical surveys. Figure 8.5, shows the occurrences of these gas hydrates all over the globe [14, 15]. Note that this representation includes both recovered and inferred gas hydrate deposits. It is clear from the representation that these hydrates are mostly present in marine sediments rather than permafrost regions. The offshore regions of North America and Russia are observed to have a lot of gas hydrate deposits.

The regions on Earth where these hydrates are likely to be formed, that is, where the external conditions are suitable for their formation, are quite a few. These include sediments at the bottom of permafrost, continental margin sedimentary deposits, deep water sediments, and under the Antarctic ice [16, 17]. Figure 8.6 shows the various types of deposits of these gas hydrates in the earth's crust and below the seabed [18]. The illustrations are not sufficient for getting a complete picture of how gas hydrates exist in the world although they give a brief overview of their presence.

Figure 8.7 shows a gas hydrate resource pyramid which depicts the relative amounts of gas hydrates globally [19]. The Arctic sands hydrate deposits, at the top of the pyramid, are likely to be used as a resource for energy in the coming decades. While moving down the pyramid, the amount of in-place gas increases while the reservoir quantity decreases. Hence, it is clear that these hydrate deposits are found all over the globe in different forms.

8.5 Indian gas hydrate scenario

India's net imports of resources like oil/coal/gas are greater than its exports. Gas hydrates can be a potential resource in India, like other countries, in the coming decades to curb this situation. There are four main sediments of gas hydrates in the Indian onshore region which are: Krishna–Godavari Basin, Mahanadi Basin, Kerala Konkan and Andaman Islands. These are clearly located in Figure 8.8 [15]. The initial recognition was done by a member from ONGC (Oil and Natural Gas Corporation Ltd) [20] who identified them through BSRs (Bottom Simulating Reflectors) in the Andaman Island Region. Over time, other surveys and investigations were conducted and gas hydrate maps were prepared [21]. Other sediments and their seismic lines, mentioned above, were identified [22]. It has been estimated that approximately 1,900 trillion cubic meters of methane gas, in the form of hydrates, are stored in the vast economic zone of India [23]. As mentioned before, this amount is nearly 1,500 times the current Indian reserve of natural gas and the utilization of a small portion of this huge quantity has the potential of solving the energy problems of India.

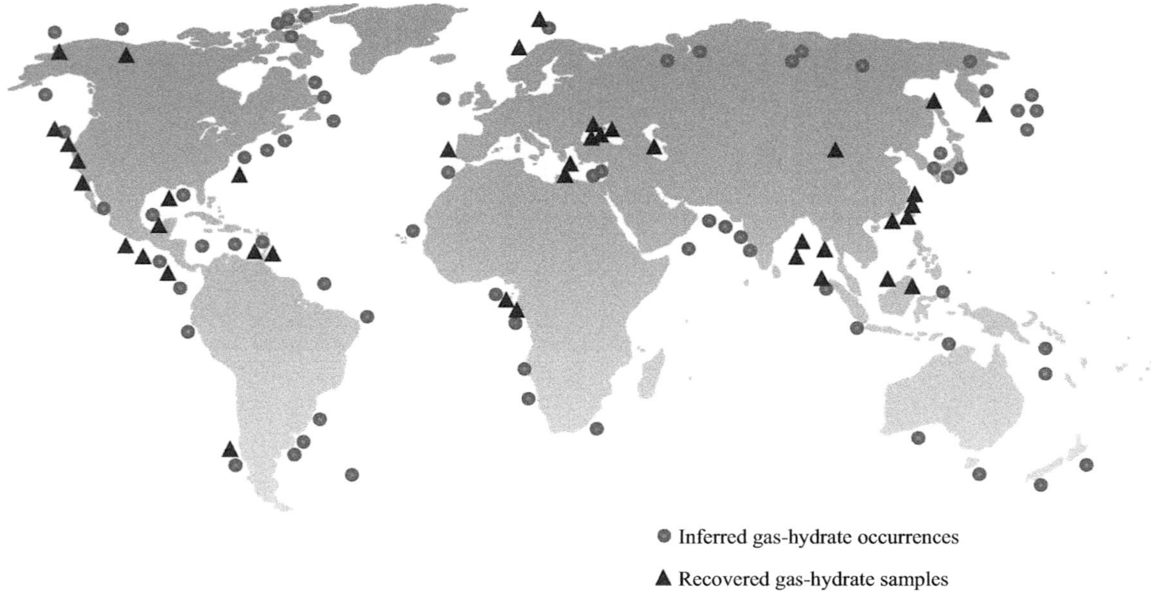

Figure 8.5 World map showing the locations of various methane hydrate deposits (Modified from References 14, 15)

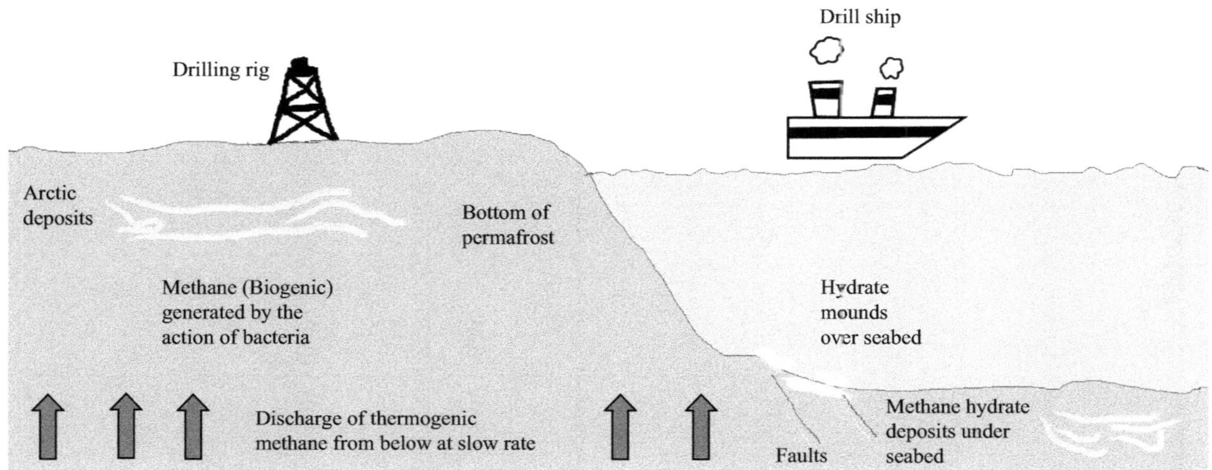

Figure 8.6 Gas hydrate deposits in permafrost and below the seabed (Modified from Reference 18)

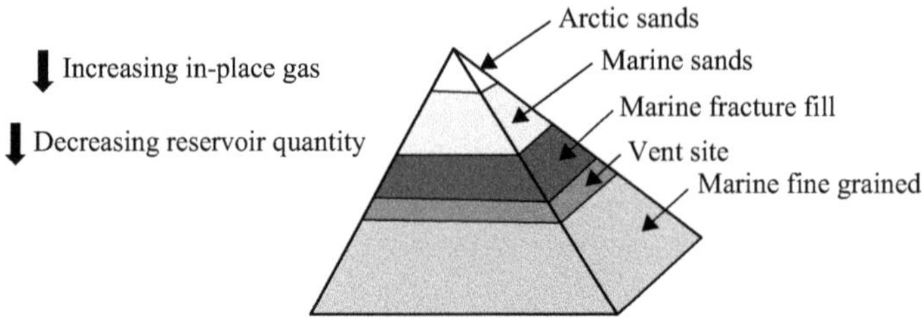

Figure 8.7 Gas hydrate resource pyramid (Modified from Reference 19)

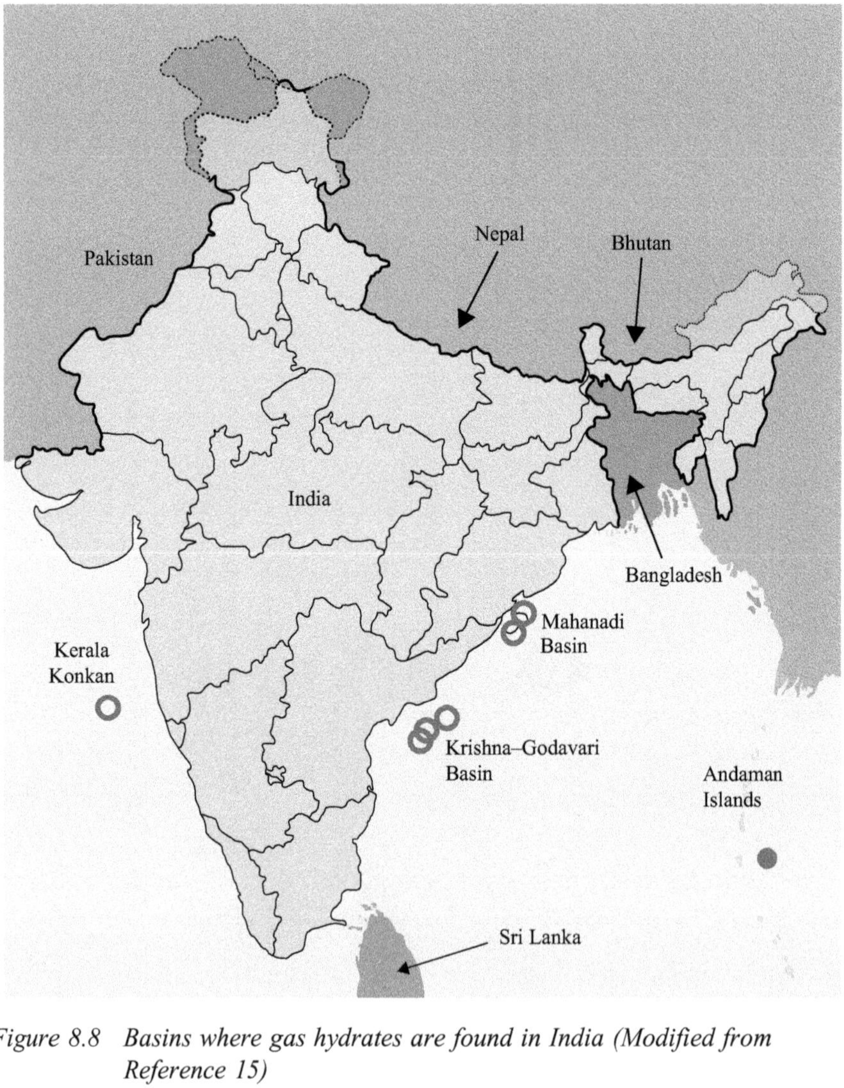

Figure 8.8 Basins where gas hydrates are found in India (Modified from Reference 15)

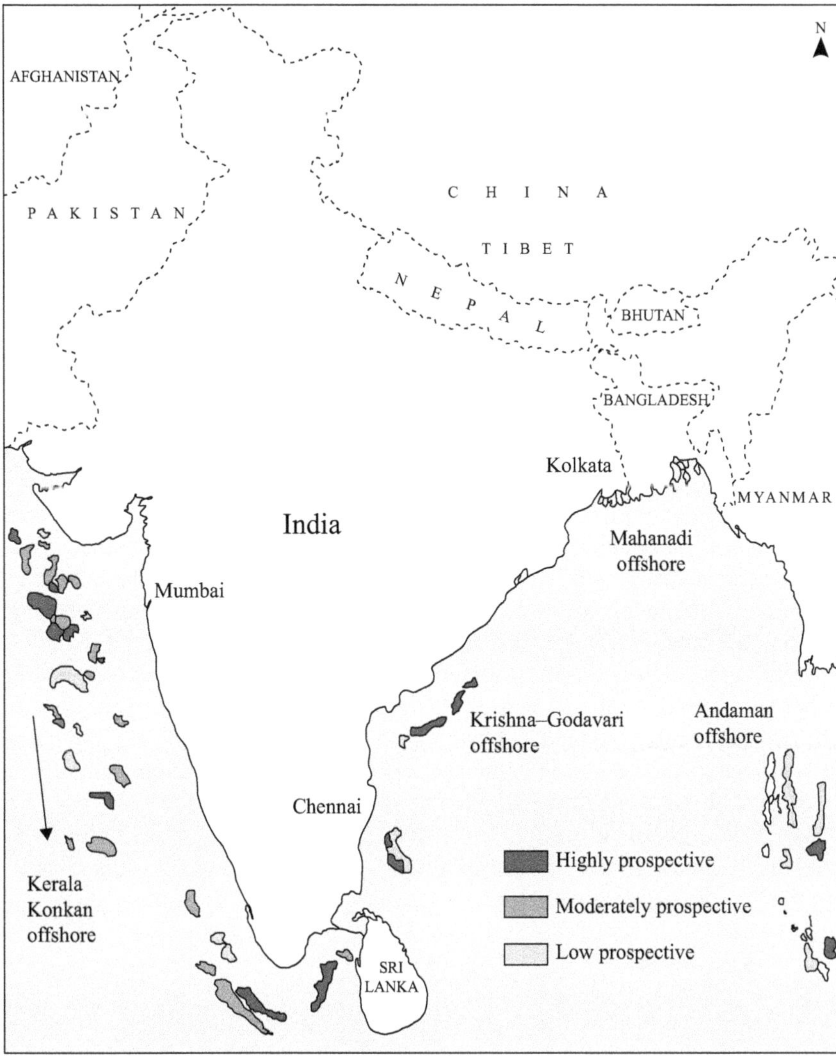

Figure 8.9 Prospective areas of gas hydrate presence in the various basins (Modified from Reference 22)

Figure 8.9 shows the prospective areas for gas hydrates along the Indian continental margins and depicts clearly the presence of various offshore deposits in the Indian subcontinental region [22].

The establishment of the NGHP took place in 1997 after its recommendation from the Ministry of Petroleum and Natural Gas (MOP&NG) expert committee, following the gas hydrate presentation in the Indian Geophysical Union (IGU) Conference held in 1995 [22]. The main agenda of this expedition was to study the occurrences of gas hydrates in the Indian peninsular region along with the

Andaman offshore region. Many sites, as mentioned before, were found at different locations that included the Kerala, Krishna–Godavari, Mahanadi and Andaman offshores [22]. The thorough details of this program can be found in other literature publications [22].

8.6 Recovery of gas hydrates

The dissociation of gas hydrates to obtain methane gas and water can be done by destabilizing or taking the hydrate out of its stability zone. This is usually done by decreasing the pressure (depressurizing) and also heating the hydrate (increasing its temperature). Since gas hydrates are formed at lower temperatures and high pressures, this provides a logical solution for methane extraction.

Figure 8.10 shows the most common extraction techniques currently used for the recovery of methane from methane gas hydrates: thermal injection, depressurization and inhibitor injection [24]. In thermal injection, hot fluid like hot water or steam is injected into the hydrate deposits through a well which causes an increase in temperature consequently destabilizing the hydrate, leading to its breakdown and the release of methane which is then collected [25, 26]. Although the rate of gas production is high in this method, it is expensive, requiring large amounts of energy to decompose small quantities of hydrate [27].

Depressurization is the most common technique used and involves releasing the pressure contained inside with the help of drilling or draining water into the wells and dissociating the methane hydrates to release methane [26, 28]. No energy source from outside is required in this method which makes it economical but it has low production rates due to the slower rate of hydrate dissociation [29].

Inhibitor injection is similar to thermal injection in which an inhibitor-like methanol, brine or glycol is injected instead of steam/hot-water. This causes a

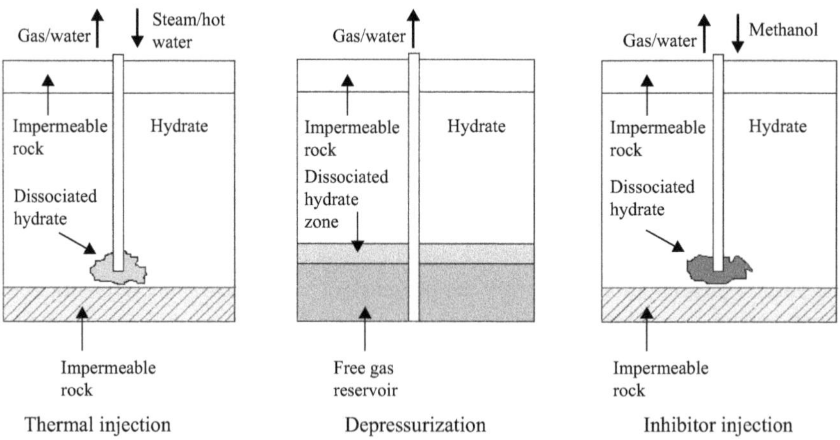

Figure 8.10 Typical methods for methane recovery (Modified from Reference 24)

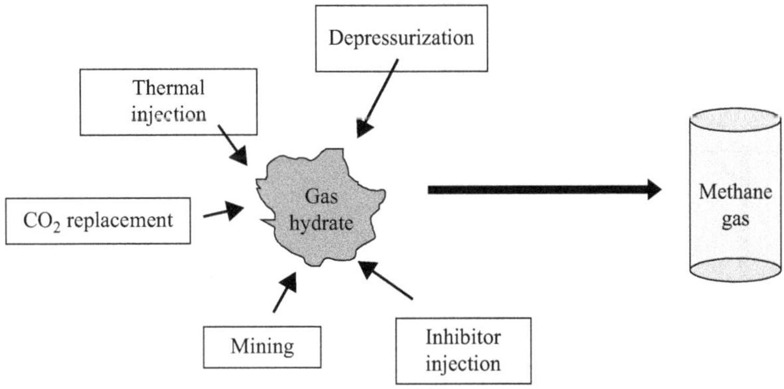

Figure 8.11 Various methods for methane recovery

destabilization of the hydrate zone. It is a comparatively expensive method since the production rate is a function of temperature, pressure of inhibitor, and the injection rate which makes it a complex process [26, 29, 30]. These methods are summarized in Figure 8.11. Thermal simulation models of these methods were reviewed recently and can be found [29].

These abovementioned methods are not the only methods for methane gas recovery. There are other methods such as carbon dioxide replacement and combinations of the above methods [26, 31]. The carbon dioxide replacement method can serve a dual purpose. It can be used for CO_2 sequestration along with methane recovery. The main pressing objective of today is to find new methods for extracting methane economically, and also to find ways to make existing methods more economical.

8.7 Environmental impacts and challenges

Some effects which come to mind when dealing with gas hydrate mining include the destabilization of seabeds. The gas hydrate deposits cement the pores present in seabeds. When these hydrates are mined, the seabeds in which they are present might become destabilized [25]. Also, there is controversial discussion relating to the occurrence of tsunami (triggering of tsunami) due to sliding of large sediments of gas hydrates downhill, although small-scale mining might not cause landslides and hence tsunami. For this reason, the carbon dioxide replacement method becomes quite desirable since it does not involve mining and does not cause destabilization of seabeds [25].

The other point that researchers discuss is the amount of methane that will be leaked into the atmosphere by the extraction of these gas hydrates. Since methane is a more potent greenhouse gas, having a global warming potential 23 times that of carbon dioxide [32], its increase would not be suggested/advisable. But in the recent edition of World Oceanic Review it was made clear that methane will not be able to reach the atmosphere if it is mined at locations where there is at least a

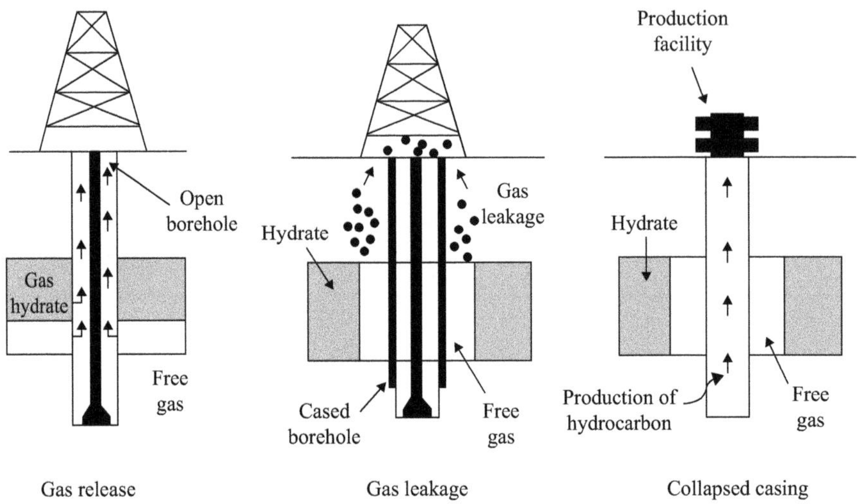

Figure 8.12 Safety issues with drilling in the Arctic. (Modified from Reference 29)

100m thick sedimentary layer over the hydrate deposits [25]. Also, methane does not shoot up on its own like other gases such as natural gas and oil do. The methane hydrate must gradually dissociate or break down, giving rise to a slow release of methane [25].

There are many challenges that may be faced while extracting and recovering hydrates. Some of these challenges include [30]:

- Unavailability of adequate technology for completion.
- Management of water production rates.
- Unavailability of representative deep water gas hydrate fields.
- Sand control of the unconsolidated sand deep below the seabeds.
- Problems relating to dissociation and evacuation.
- Ultimate recovery and sea floor subsidence.

The abovementioned points are not all of the issues concerning the extraction of gas hydrates. Another issue includes problems while drilling in arctic and marine settings. The issues associated with drilling in arctic settings are shown in Figure 8.12 [29].

Gas release is caused when penetration by the drill bit causes dissociation of gas hydrates underneath and gas then moves up the open hole. Gas leakage and collapsing of casing occurs when the region of gas hydrate intervals are cased or cemented [33].

8.8 Using hydrates for methane storage

Due to the metastable nature of hydrates at atmospheric pressure and temperatures which are slightly lower than the ice point, researchers are investigating how hydrates can be used to store and transport methane [34]. Hydrates have the potential of being not only an abundant source of methane, but also a means of methane storage.

Various methods of enhancing the rate of methane hydrate formation and storage capacity of methane hydrates have been considered, including the use of surfactants, silica, activated carbon and amino acids [35–40]. These methods may improve the feasibility of methane hydrates to be used to store methane for transportation or for the use in applications such as natural gas vehicles [40].

8.9 Future perspective and conclusion

The dependency of the current generation on fossil fuels for most of its energy needs suggests that switching to renewable sources might not be as simple as expected. It will take time to make solar, wind, tidal, biomass etc. the main sources of energy. Since fossil fuel reserves are not sufficient to last long in many parts of the world, gas hydrates, as discussed above, will play an important role in solving major energy problems of developing and developed nations. Figure 8.13, from the US Geological Survey [41], shows the roadmap of natural gas hydrates consisting of history and predicted events that are expected to occur in the near future. It can be seen from the figure that many more field trials are estimated to be done for

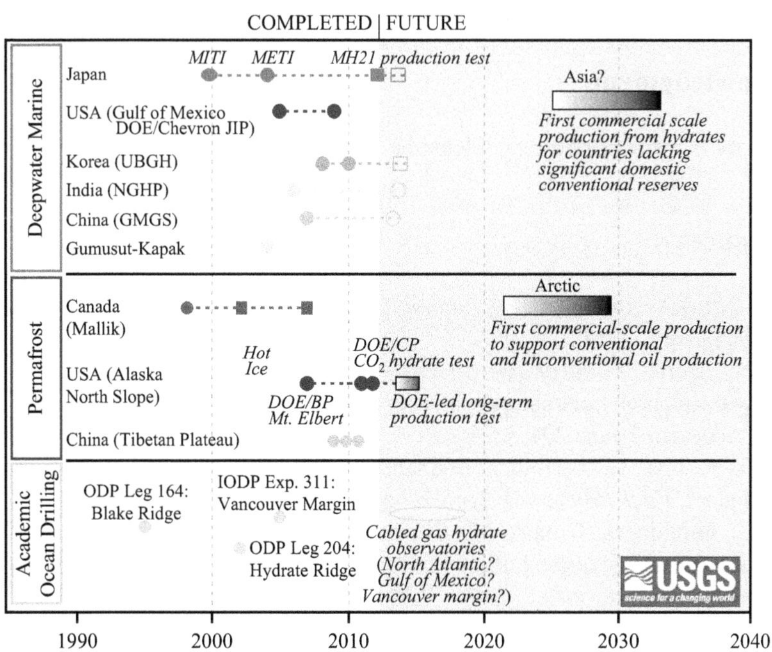

Figure 8.13 Roadmap for natural gas hydrates (Adopted from USGS [41])
Credit: U.S. Geological Survey Department of the Interior/USGS

long-term production which will be essential for the realization of commercial production from these hydrates in the coming three decades. It is predicted that the results and experiences obtained from these tests and previous tests will lead to the first commercial scale production in the Canadian/US region in the mid-2020s. Asia might be the region for the first commercial scale offshore production, most probably in Japan or Korea. The figure suggests that a significant level of natural gas hydrate production is expected after the 2030s [41]. A good amount of research work and experiments might shift these times nearer to the present time, which will be a boost to economies and GDPs of some nations.

In conclusion, there is considerable availability of methane gas hydrates all over the globe. There are various methods available that can be used for recovery of methane from its hydrates which include depressurization, thermal injection, inhibitor injection, mining and carbon dioxide replacement, but these processes might not be viable for the commercial scale production of these hydrates. It has been estimated that it will take two decades until commercial scale production will be possible. As far as India is concerned, it also has significant potential for development in the field of gas hydrates and synthesizing and utilization of even a small amount of these hydrates will solve major energy problems. The coming few decades may see gas hydrates become a major source for energy production, walking hand-in-hand with renewable sources of energy. Additionally, studies into improving the rate of methane hydrate formation and methane hydrate storage capacity will also enable their use for methane storage.

Acknowledgment

This work is made possible by Globalink Mitacs and NSERC.

References

[1] US E.I.A. *Annual energy outlook 2013*, Washington, DC: U.S. Department of Energy, 2014.

[2] N. Laurell. *Natural gas overview: why methane is cleaner fuel*? http://www.nlaurell.com/natural-gas-overview-why-is-methane-a-clean-fuel/, 2014 [Accessed 2 June 2015].

[3] S.-Y. Lee, G.D. Holder. Methane hydrates potential as a future energy source, *Fuel Processing Technology* 71: 181–186, 2001.

[4] C. Giraldo, J. Klump, M. Clarke, J.M. Schicks. Sensitivity analysis of parameters governing the recovery of methane from natural gas hydrate reservoirs, *Energies* 7: 2148–2176, 2014.

[5] S. Adisasmito, R.J. Frank III, E.D. Sloan Jr. Hydrates of carbon dioxide and methane mixtures, *Journal of Chemical & Engineering Data* 36: 68–71, 1991.

[6] T. Collett, J.-J. Bahk, R. Baker, et al. Methane hydrates in nature – current knowledge and challenges, *Journal of Chemical & Engineering Data* 60: 319–329, 2015.

[7] Z.R. Chong, S.H.B. Yang, P. Babu, P. Linga, X.S. Li. Review of natural gas hydrates as an energy resource: prospects and challenges, *Applied Energy*, 162, 1633–1652, 2016.

[8] K. Sain, H. Gupta. Gas hydrates in India: potential and development, *Gondwana Research*, 22(2): 645–657, 2012.

[9] S.A. Nelson, *Coal and gas hydrate resources*, Tulane University, 2006 http://earthsci.org/education/teacher/basicgeol/coal/coal.html [Accessed 28 May 2015].

[10] K.A. Kvenvolden. A review of the geochemistry of methane in natural gas hydrate, *Organic Geochemistry* 23(11/12): 997–1008, 1995.

[11] Y.F. Makogon. Natural gas-hydrates—a potential energy source for the 21st century, *Journal of Petroleum Science and Engineering* 56: 14–31, 2007.

[12] Y.F. Makogon. *Hydrates of hydrocarbons*, Tulsa: PennWell Publishing Co., 1997.

[13] E.D. Sloan. *Clathrate hydrates of natural gases*, 3rd Edition, Taylor and Francis Group: CRC Press, 2008.

[14] S. Lu. A global survey of gas hydrate development and reserves: specifically in the marine field, *Renewable and Sustainable Energy Reviews*, 41: 884–900, 2015.

[15] R. Birchwood, J. Dai, D. Shelander, *et al.* Development in gas hydrates, *Oilfield Review*, 22(1): 18–33, 2010.

[16] H. King. *Methane hydrate*, http://geology.com/articles/methane-hydrates/ [Accessed 14 May 2015].

[17] T. Stephens. *Study suggests large methane reservoirs beneath Antarctic ice sheet: press release*, Santa Cruz: University of California, August 29, 2012.

[18] *Potential of gas hydrates is great, but practical development is far off*, 2012 http://www.eia.gov/todayinenergy/detail.cfm?id=8690# [Accessed 12 May 2015].

[19] R. Boswell. *The gas hydrate resource pyramid, fire in the ice*, Methane Hydrate Newsletter, Natural Gas & Oil 304: 285–4541, 2006.

[20] N.N. Chopra. Gas hydrate an unconventional trap in forearc regions of the Andaman offshore, *Bulletin of the Oil and Natural Gas Commission*, 22 (1): 41–54, 1985.

[21] Y.H. Rao. *Gas hydrate investigations along the continental margins of India*, PhD dissertation, Hyderabad, India: Osmania University, 2001.

[22] T.S. Collett, R. Boswell, J.R. Cochran, *et al.*, NGHP Expedition 01 scientific party, geologic implications of gas hydrates in the offshore of India: results of the National Gas Hydrate Program Expedition 01, *Marine and Petroleum Geology*, 58(A): 3–28, 2014.

[23] T.S. Collet, M. Riedel, J. Cochran, *et al.*, NGHP Expedition 01 Scientists, 2008. NGHP Expedition 01, Initial Reports, Directorate General of

Hydrocarbons, Noida and Ministry of Petroleum & Natural Gas, India, vol. 4, 2006.
[24] G. Ersland, A. Grauce. Natural gas hydrates, in *Natural Gas* edited by P. Potocnik, InTech, 2010, http://www.intechopen.com/books/natural-gas/natural-gas-hydrates [Accessed 1 June 2015].
[25] World Ocean Review, *WOR 3*, maribus gGmbH, 2014.
[26] Y. Song, C. Cheng, J. Zhao, *et al.* Evaluation of gas production from methane hydrates using depressurization, thermal stimulation and combined methods, *Applied Energy* 145: 265–277, 2015.
[27] DGH India, *Indian National Gas Hydrate Program*, http://www.ccop.or.th/download/PETRAD/PETRAD59_2011–03/Doc-10_MalcolmLall.pdf [Accessed 25 May 2015].
[28] *MH21 research consortium*, 2008 http://www.mh21japan.gr.jp/english/mh21–1/3–2/ [Accessed 18 May 2015].
[29] Y. Sun, X. Lu, W. Guo. A review on simulation models for exploration and exploitation of natural gas hydrate, *Arabian Journal of Geosciences*, 7(6): 2199–2214, 2014.
[30] J. Chen, J. Liu, G. Chen, *et al.* Insights into methane hydrate formation, agglomeration, and dissociation in water + diesel oil dispersed system, *Energy Conversion and Management*, 86: 886–891, 2014.
[31] P. Bhade, J. Phirani. Gas production from layered methane hydrate reservoirs, *Energy*, 82: 686–696, 2015.
[32] C.D. Goldsmith Jr., J. Chanton, T. Abichou, N. Swan, R. Green, G. Hater. Methane emissions from 20 landfills across the United States using vertical radial plume mapping, *Journal of Air Waste and Management* 62(2): 183–197, 2012.
[33] "Energy from Gas Hydrates – Assessing the Opportunities and Challenges for Canada Report of the Expert Panel on Gas Hydrates," 2008.
[34] H. Ganji, M. Manteghian, K. Sadaghiani zadeh, M.R. Omidkhah, H. Rahimi Mofrad. Effect of different surfactants on methane hydrate formation rate, stability and storage capacity, *Fuel* 86(3): 434–441, 2007.
[35] V.D. Chari, D.V.S.G.K. Sharma, R.S.R. Prasad, S.R. Murthy. Methane hydrates formation and dissociation in nano silica suspension, *Journal of Natural Gas Science and Engineering* 11: 7–11, 2013.
[36] P.S.R. Prasad, Y. Sowjanya, V.D. Chari. Enhancement in methane storage capacity in gas hydrates formed in hollow silica, *Journal of Physical Chemistry C* 118(15): 7759–7764, 2014.
[37] S.-Q. Hao, S. Kim, Y. Qin, X.-H. Fu. Enhanced methane hydrate storage using sodium dodecyl sulfate and coal, *Environmental Chemistry Letters* 12(2): 341–346, 2014.
[38] B. Liu, W. Wang, N. Wang, C.T. Au. Preparation of activated carbon with high surface area for high-capacity methane storage, *Journal of Energy Chemistry* 23(5): 662–668, 2014.

[39] A. Fazlali, S.-A. Kazemi, M. Keshavarz-Moraveji, A.H. Mohammadi. Impact of different surfactants and their mixtures on methane-hydrate formation, *Energy Technology* 1(8): 471–477, 2013.
[40] Y. Liu, B. Chen, Y. Chen, *et al.* Methane storage in a hydrated form as promoted by leucines for possible application to natural gas transportation and storage, *Energy Technology* 3(8): 815–819, 2015.
[41] *U.S. geological survey gas hydrates project*, http://woodshole.er.usgs.gov/project-pages/hydrates/primer.html [Accessed 5 June 2015].

Index

www.ingramcontent.com/pod-product-compliance
Ingram Content Group UK Ltd.
Pitfield, Milton Keynes, MK11 3LW, UK
UKHW020820290926
15470UKWH00003B/1282

9 781785 611933